NORTH CAR
STATE BOARD OF COM
LIBRAR
ASHEVILLE-BUNCOMBE TECHNICAL COMMUNITY COLLEGE

HUMIDITY CONTROL

A humidification guide for
residential, commercial and
industrial applications

Ivan Stepnich

DISCARDED

DEC 1 0 2024

BNP Business News Publishing Company
Troy, Michigan

Copyright © 1988
Ivan Stepnich

All rights reserved. No part of this book may be reproduced or transmitted in any form or by any means—electronic or mechanical, including photocopying, recording or by any information storage and retrieval system—without written permission from the publisher, Business News Publishing Company.

Administrative Editor: Phillip R. Roman

Library of Congress Cataloging in Publication Data
Stepnich, Ivan C.
    Humidity control: a humidification guide for residential, commercial and industrial applications/Ivan Stepnich.
      p. cm.
   ISBN 0-912524-48-0
   1. Humidity—Control. I. Title.
TH7687.8.S74   1988
697.9'323—dc19   88-39202
                                            CIP

CONTENTS

Introduction

Section 1   Humidity Fundamentals 1
Section 2   Psychrometric Charts 16
Section 3   Residential Humidification 42
Section 4   Residential Installation & Service 63
Section 5   Commercial, Industrial,
            Institutional Humidification 93
Section 6   Dehumidification 129

Index 153

DISCLAIMER

This book is only considered to be a general guide. The author and publisher have neither liability nor can they be responsible to any person or entity for any misunderstanding, misuse or misapplication that would cause loss or damage of any kind, including material or personal injury, or alleged to be caused directly or indirectly by the information contained in this book.

PREFACE

The focus of this text is to assist and guide individuals in properly designing, selecting, installing and maintaining the humidification/dehumidification equipment necessarily required in the environmental control systems of various types of buildings.

Humidity control has come a long way since the days when folks merely placed a pot of water on a hot radiator to increase the humidity in a room. However, for all practical purposes, control of relative humidity is still not overly complicated. It is essential that everyone from the designer to the installer (and even the occupants of a building) become cognizant of the numerous benefits of humidification. Although contrary to common belief, these benefits do not apply solely to residential living environments. By controlling the humidity of any building, occupants gain a cleaner, more comfortable environment which fosters better health. One study of twelve schools indicated that absenteeism due to illness was significantly decreased when the relative humidity was increased from 20% to 35%. Furthermore, in commercial and industrial settings, scientific studies have proven that properly humidified work sites tend to induce increased productivity which can result in labor savings as well as more efficient equipment operation.

INTRODUCTION

If scientists from other parts of the universe could focus on planet earth, their instant reaction would be —water, everywhere water. Water not only covers ¾ of the earth's total surface area; in vapor form, water blankets all land masses and is a major factor in weather phenomena.

In fact, our water planet has exerted a profound influence on all forms of life throughout the evolutionary process. The human body, for example, is primarily composed of water. It is highly dependent on water evaporation to maintain the delicate balance between net heat produced by body metabolism and net heat lost to the environment.

Scientists have developed highly complex formulas to define physiological comfort. And although these formulas may be highly speculative, the following key values are well established and include:
- temperature
- humidity
- air motion
- air purity
- sound

The most important yet least understood value is humidity. Despite this, humidification equipment is rarely designed into original heating or environmental control equipment. Instead it is relegated as an "after-market" item which must be special ordered. This means that most humidifier applications must be engineered and installed at the trade level. The net result is that some applications are properly engineered, installed and serviced. However, many are not. In academic circles, economics is called the "dismal science" because of the inexpertness of its practitioners. For exactly the same reason humidification should be called the "dismal technology." This does not reflect on the technical competence of field personnel. Rather, it reflects inadequate technical information and training procedures related to humidification.

Humidifiers are not complex—contrary to common opinion. Engineering, application and servicing, however, do require a solid understanding of basic principles and procedures. Consequently, the purpose of this book is to provide those guidelines in order to foster well-engineered, effective and profitable humidifier installations.

SECTION 1

HUMIDITY FUNDAMENTALS

ENERGY AND MATTER

Humidifying basically involves heat energy and its interaction with water. Heat energy conforms to a very fundamental law of thermodynamics—heat always flows from a higher to a lower source. For example, consider how water flows. Just as water flows downhill, heat must also flow downhill or to an area with lesser heat.

Heat Units

The unit which expresses the measurement of heat energy is the British Thermal Unit (Btu). One Btu is equivalent to the amount of heat required to raise one pound of water one degree Fahrenheit. On the other hand, if a single Btu is extracted from one pound of water, the temperature falls one degree Fahrenheit. Consequently, the assigned value of a Btu is based on the specific heat of water.

Heat power or heat rate is expressed in Btuh (Btu per hour). In metric terminology, the counter part of the Btu is the calorie. By definition, a calorie is the amount of heat required to raise one gram of water one degree centigrade.

Most foreign countries express heat units as calories. However, the Btu is still the established unit of heat measurement observed in the United States.

Heat Measurement

A thermometer is an instrument that indicates just how hot a substance is—not its heat quantity. Consequently, a thermometer merely measures heat intensity. Ten pounds of water at a temperature of 200°F has a much higher temperature than would 100 pounds of water at 100°F. Nevertheless, the water at 100°F has more than four times as much total heat. Total heat of a substance depends upon: specific heat, temperature and mass (measured in pounds or grams).

Note: The specific heat of a substance is the amount of heat required to raise the temperature of one pound 1°F.

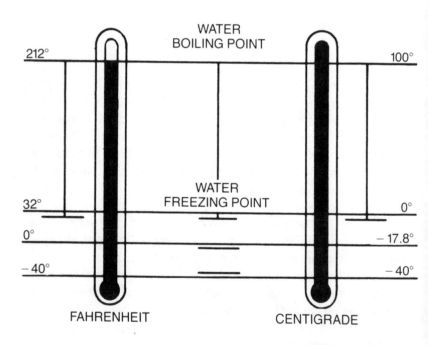

Figure 1-1. Illustration depicting the comparison between the Centigrade and Fahrenheit temperature scales.

Fundamentals

As noted previously, the specific heat of water is 1 Btu/lb; the specific heat of dry air is 0.24 Btu/lb.

A comparative illustration of the Fahrenheit and centigrade (metric) thermometers is shown in Figure 1-1. According to the principles of physics, all matter has heat energy starting from absolute zero temperature. Absolute zero is $-460°$ below $0°F$ and $-273°$ below $0°C$. To get the highest accuracy, all mathematical calculations involving temperature must be converted to the absolute zero temperature scale. This conversion is accomplished by adding 460 to the Fahrenheit temperature reading or 273 to the centigrade reading.

On the Fahrenheit scale, water freezes at $32°$ and boils (at sea level) at $212°$. Zero on the centigrade scale is equivalent to the freezing point of water on the Fahrenheit scale ($32°$), and $100°$ ($212°F$) indicates the boiling point of water at sea level.

HEAT TRANSFER

Heat travels from a higher energy level to a lower energy level. The rate at which heat travels, however, depends on the following:
- mass or size of heated substance
- specific heat of the substance
- temperature difference

Three methods of heat transmission exist: conduction, convection and radiation. Each are illustrated in Figure 1-2. Most humidifiers utilize any one or a combination of these methods in transferring heat energy to vaporized water.
- Conduction — If one end of an iron rod is heated, the opposite end eventually gets hot as well. Heat is conveyed from one end of the rod to the other by the process of conduction which involves the molecule-by-molecule transmission of heat. When an electrical heating element is immersed in a humidifier reservoir, the process of conduction transfers heat from the element directly to the water.
- Convection — As water in the bottom of the humidifier reservoir is heated by the immersion element, the heated water

has greater energy and becomes lighter in weight due to molecular expansion. Heated water rises towards the surface of the reservoir because it is lighter than the cooler water that surrounds it. As the heated water rises, it gives up its heat to the cooler surrounding water (i.e., loses its heat energy), becomes heavier and returns to the bottom of the reservoir where the cycle is repeated. This continuous process of heat transfer by a moving fluid such as air or water is known as convection.

- **Radiation** — Heat radiation is emitted from all bodies and all surfaces which exist at temperatures above absolute zero. The higher the temperature, the greater the emission. In fact, infrared humidification processes rely on direct radiation to vaporize water. The use of infrared heat sources simulates solar heat radiation which tends to vaporize natural bodies of water such as rivers, lakes and oceans.

    The advantage of radiation, energy is transmitted at the speed of light and the heat source does not require physical contact with the water. On the other hand, for convection or conduction to occur, the heat energy source must have direct physical contact with the water. This results in a pronounced lag-time before any energy transfer occurs and a continuation or "fly-wheel" effect when heat energy is stopped.

CHANGE OF STATE

When a substance absorbs heat, its molecules increase in motion as temperature increases. For example, as ice absorbs heat, its molecular motion increases in direct proportion to the temperature increase. This continues until ice reaches the temperature of 32°F (0°C). Although it takes one Btu to raise one pound of water one degree, the specific heat of ice is somewhat less. It only takes 0.48 Btu to raise one pound of ice one °F.

Fundamentals

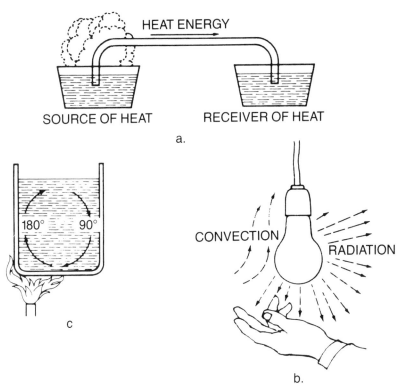

Figure 1-2. Three methods of heat transmission: a) conduction, b) radiation, c) convection — convection current depicted.

Sensible Heat

Any heat energy that can be detected by a thermometer is considered sensible heat. It derives its name in that it can be perceived by the senses. For example, through a sense of touch, it becomes obvious that an object is getting warmer. In short, the increased molecular activity caused by applying heat energy can be sensed directly.

Latent Heat of Fusion

After ice absorbs enough heat to reach 32°F, it continues to absorb heat, but its temperature does not increase. Instead

Humidity Control

this added heat energy causes the molecules to push farther and farther apart. When 144 Btu/lb have been absorbed, this repulsion causes the ice to change state (melt) with no discernable increase in temperature. The 144 Btu which caused 32°F ice to turn into 32°F water is known as the latent heat of fusion. The term latent is used because it means hidden or not perceptible to the senses. Since the latent heat of fusion causes a change of state with no increase in temperature, it is not apparent to the senses and cannot be measured with a thermometer. The 144 Btu will only be given up to the surrounding air when water changes back into ice. Water like most substances can exist as a solid (ice), a liquid and a gas (vapor). Figure 1-3 illustrates this change of state phenomena.

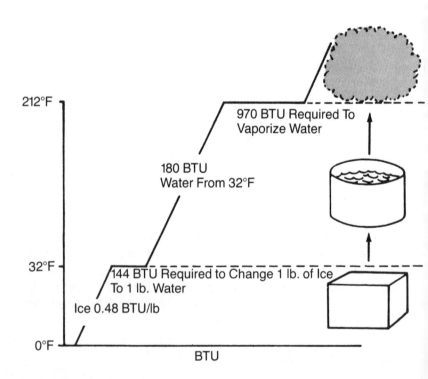

Figure 1-3. Change of state process showing heat requirements for conversions.

Fundamentals

> Note: The amount of heat energy absorbed by a substance determines its state.

Latent Heat of Vaporization

As water continues to absorb heat, it increases in temperature at the rate of 1°F/Btu or 1°C/calorie until it reaches the boiling point. At sea level pressure this is equivalent to 212°F or 100°C. If additional heat is applied after water reaches the boiling point, velocity of the molecules does not increase. Instead, the additional heat forces the molecules farther apart and causes the molecules to escape into the atmosphere as a true gas. At sea level pressure it takes 970 Btu to change one pound of water at 212°F into one pound of steam at 212°F. The 970 Btu required to convert water from its liquid state into the gaseous (vapor) state is referred to as the latent heat of vaporization.

Again the 970 Btu required to convert water into vapor is latent heat. Remember, latent heat is not perceptible to the senses and is only given up (released) when the vapor condenses back into water.

The water vapor can be heated beyond 212°F, however, and the result is known as superheated vapor. The specific heat of water vapor is 0.48 Btu/lb. Consequently, 0.48 Btu is required to raise the temperature of one pound of vapor 1°F.

Liquids have the ability to absorb or reject relatively large quantities of heat when changing state. In humans, for example, body temperature is controlled by the vaporization of perspiration. A relatively small quantity of perspiration can dissipate a great deal of body heat in the temperature regulating process. The ability of liquids to absorb or reject heat is the phenomenon which also makes refrigeration practical. In a refrigeration or air conditioning unit, the vaporizing liquid refrigerant can absorb large quantities of heat where it is not needed, and at the appropriate stage, release it into the atmosphere by condensing the vapor back into a liquid.

Heat and Humidity

Even though water has the ability to absorb large quantities of heat energy, it is not a significant advantage in humidification. About 970 Btu are required to vaporize a pound of water at its boiling point (212°F). In order to vaporize water at temperatures lower than the boiling point (212°F), additional heat is required. At a vaporizing temperature of 140°F, for example, approximately 1,014 Btu are necessary to vaporize a pound of water. However, water can vaporize at any temperature where the vapor pressure is low enough to allow the molecules to escape. In fact, even ice can be converted directly into vapor by a process known as sublimation.

In terms of commonly used heating fuels, vaporizing a pound of water requires approximately the heat energy of one cubic foot of natural gas, 300 watts of electricity or an ounce of fuel oil. If coal is used as a heat source, approximately $1/12$ pound is necessary to vaporize a pound of water. Although the amount of fuel required to vaporize just one pound of water may seem like a lot, this heat energy is not wasted. The heat energy provides a comfort level at a lower thermostat setting which appreciably reduces a building's heat loss and conserves energy. If the building is a typical pre-World War II model, however, which lacks vapor barriers and has drafty windows and doors, vapor escapes readily and little energy is conserved by humidifying. In fact, humidifying such a house can actually increase energy waste. In buildings constructed with properly sealed/weather-stripped doors and windows, dampered fireplaces and adequate vapor barriers, energy waste should be minimal. In modern buildings humidification improves comfort and promotes a healthier living environment—while conserving energy.

Fundamentals

HUMIDITY AND GAS PROPERTIES

Air in our atmosphere is a mixture of many gaseous elements: nitrogen, oxygen, argon, carbon dioxide and water vapor. There are also small traces of hydrogen, helium, krypton, xenon and neon. Despite the minute presence of rare gases, air primarily consists of oxygen, nitrogen and water vapor. Consequently, dry air is basically oxygen and nitrogen. Since these two gases are consistently found in the same proportions by volume (79% nitrogen and 21% oxygen), they are considered a single gas for the purpose of humidification and air-conditioning calculations.

Water vapor, on the other hand, is the variable constituent in air mixtures. In 1,000 ft^3 of air at subzero temperatures as little as .005 lb of water vapor can be present. Comparatively, air in the 100°F range can hold as much as 3 pounds of water vapor.

Despite this fact, water vapor is a true gas and is always present in air in varying amounts. As a gas, water vapor completely loses its identity and individuality as a substance and is governed exclusively by the universal gas laws. One key characteristic of gas is that it automatically fills any container—regardless of size. Liquids and solids lack this ability. The following are some of the laws governing the expansion and contraction of gases:

- Boyle's Law—If temperature is constant, the volume of a gas varies inversely with pressure. If pressure increases, volume decreases.
- Charles' Law—If pressure is constant, the absolute volume varies with the absolute temperature. If temperature increases, volume increases proportionally.
- The Universal Gas Formula: $PV = MRT$ This formula address each variable and is used to determine the basic relationships of a gas. The values of the formula:
  P = Pressure exerted by a gas (lb/ft^2)
  V = Volume of gas (ft^3)
  M = Weight of gas (lb)
  R = Gas constant (dry air = 53.35)
  T = Absolute temperature

Humidity Control

The constant for any gas (R) is derived by dividing 1,546 by the molecular weight of the gas. Note, values for temperature and pressure must be absolute values. The value for pressure also must be in lb/ft² absolute instead of lbs/in².

> Note: The gas formula is calculated in lbs/ft², not lbs/in² (psi).

- **Dalton's Law of Partial Pressures** — The other gas laws may be useful in mathematically calculating basic factors such as volume and pressure with varying temperatures. Those calculations, however, are seldom required in conventional humidification applications. Dalton's law, on the other hand, involves no calculation but provides a greater insight into the behavior and distribution of humidity than do the other gas laws. Simply stated, Dalton's gas law is as follows:

    When a mixture of gases is confined in an area, each gas exerts a partial pressure equal to the pressure it exerts if occupying the space alone.

    Understand, gas expands to fill its enclosure. Although the molecules are in rapid motion (2,000 ft/second), the expansion results in sufficient space to provide a minimum of molecular interference. This means that the total pressure in a confined area is the sum of all the partial pressures from all gases within the area.

    Figure 1-4 graphically depicts the important factors in Dalton's Law, but it does not explain how it affects the humidification of a building. According to the principles of Dalton's Law, vapor migrates in a linear manner at the velocity of 2,000 ft/second. Vapor can only be confined by a restraining barrier. The restraining barrier, however, must be more than just ordinary wall construction. It must be wall construction with the ability to confine a gas. Otherwise, vapor can penetrate ordinary construction and pass through. When this happens to interior wall partitions, adjacent rooms gain humidity. When vapor passes through walls to the outdoors

Fundamentals

(exfiltrates), it creates a load requirement which must be compensated by humidification equipment. The equipment must expend about 1,000 Btu/lb of vapor to overcome exfiltration loss.

Unfortunately, the migration of water vapor is not easily discernable because pure water vapor is colorless and odorless. If water vapor had the odor of boiling cabbage, its ability to migrate in all directions and to the exterior of poorly constructed buildings could be quickly recognized. Although it might sound silly, it would at least serve as a reminder of the important role vapor barriers play in building quality.

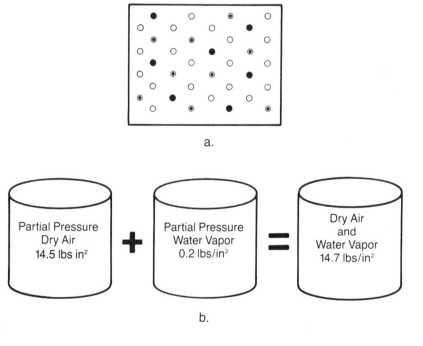

Figure 1-4. a) Daltons Law: different gases may occupy the same space. However, each gas exerts its own pressure as if the other gases were not present, b) result when the partial pressures of water vapor and dry air are combined.

RELATIVE HUMIDITY

Relative humidity is the most familiar term associated with humidity and humidification. The American Society of Heating, Refrigerating and Air-Conditioning Engineers (ASHRAE), defines relative humidity as the ratio of the mole fraction of water vapor (in a given moist air sample) to the mole fraction in an saturated air sample which at the same temperature. A mole is the gram molecular weight of vapor. This definition may be scientifically accurate, but it does not readily lend itself to practical applications.

Relative Humidity Example

As previously indicated, water vapor is the most extreme variable in an air mixture. The amount of vapor that air can hold varies significantly with temperature. Table 1-1 shows the total amount of vapor that 1,000 ft^3 can hold at various temperatures and relative humidities. Reading to the right from the 0°F line reveals 0.059 lbs of vapor at saturation or 100% relative humidity (abbreviated RH) in 1,000 ft^3. At 72°F the same 1,000 ft^3 can hold 1.273 lbs at saturation. Consequently, the saturated 0° air heated to 72°F actually has less than 5% RH.

Definition Relative Humidity

The true relative humidity in the foregoing example is 4.8%. It was derived by using the vapor pressure of water at 0°F as compared to the vapor pressure of water at 72°F. Vapor pressure is the best value for accurately computing relative humidity. The reason, vapor pressure rather than vapor quantity determines the speed of evaporation. Therefore, the following is a technically accurate and understandable definition of relative humidity:

Relative humidity is the water vapor pressure actually present in the air divided by the vapor pressure when the air is throughly saturated with vapor.

Fundamentals

Table 1-1. Total amount of vapor which 1,000 ft³ can hold at various temperatures and relative humidities.

Moisture Content*   % Relative Humidity

Temp. °F	20	25	30	35	40	45	50	55	60	70	80	90	100
-10	.008	.010	.012	.014	.016	.018	.020	.022	.024	.028	.032	.036	.040
0	.012	.015	.018	.021	.024	.027	.029	.032	.035	.041	.047	.052	.059
10	.020	.025	.029	.034	.039	.044	.049	.054	.059	.069	.078	.088	.098
20	.031	.039	.047	.055	.063	.071	.079	.087	.095	.110	.126	.142	.161
30	.051	.064	.077	.089	.103	.116	.128	.141	.154	.180	.203	.232	.258
40	.077	.097	.117	.135	.155	.174	.193	.213	.233	.272	.310	.358	.390
50	.123	.141	.171	.198	.227	.259	.284	.312	.341	.399	.456	.521	.574
60	.162	.205	.246	.288	.329	.370	.412	.453	.495	.579	.663	.747	.831
62	.176	.220	.264	.308	.353	.398	.442	.486	.532	.622	.711	.801	.893
64	.189	.236	.284	.332	.379	.427	.475	.524	.572	.668	.764	.863	.960
66	.203	.254	.306	.356	.407	.458	.510	.561	.613	.717	.821	.926	1.030
68	.217	.272	.326	.381	.436	.491	.547	.602	.657	.768	.880	.991	1.106
70	.232	.291	.349	.408	.467	.527	.585	.645	.704	.824	.944	1.063	1.186
72	.249	.312	.375	.438	.501	.564	.628	.691	.755	.884	1.013	1.139	1.273
74	.267	.334	.401	.469	.536	.605	.671	.740	.809	.947	1.084	1.224	1.364
76	.285	.357	.429	.502	.574	.647	.720	.792	.866	1.012	1.160	1.310	1.461
78	.305	.382	.458	.535	.614	.692	.769	.848	.925	1.082	1.241	1.400	1.564
80	.326	.408	.490	.574	.656	.740	.823	.905	.991	1.160	1.330	1.500	1.676
82	.348	.436	.524	.612	.701	.790	.881	.970	1.060	1.240	1.422	1.609	1.792
84	.371	.464	.558	.654	.749	.844	.939	1.033	1.129	1.322	1.520	1.710	1.916
86	.389	.496	.597	.698	.799	.901	1.003	1.105	1.212	1.418	1.627	1.833	2.048
88	.422	.523	.635	.743	.851	.960	1.068	1.178	1.288	1.510	1.731	1.958	2.189
90	.450	.563	.678	.793	.908	1.024	1.141	1.258	1.375	1.613	1.853	2.095	2.338

*Pounds per 1000 cubic feet of air (based on standard conditions)

Vapor pressures corresponding to the given temperatures can be found in the tables know as, **Properties of Water at Saturation**. These tables are published in the *ASHRAE Handbook of Fundamentals* as well as in other industry handbooks. Besides being presented in tables, vapor pressures are also shown in the psychrometric charts. The vapor pressure units in the charts are sometimes expressed in inches of mercury (Hg) instead of lbs, but this does not affect the relative humidity calculations. The psychrometric chart will be addressed in an ensuing section.

PERCENTAGE HUMIDITY

Percentage humidity is not difficult to understand since it is expressed in terms of actual pounds of water vapor present in room air compared to the potential maximum number of pounds the air is capable of of holding when fully saturated. This figure is determined by dividing the weight of vapor present by the given weight of vapor one lb of dry air can potentially hold. In a psychrometric chart these values are shown as the humidity ratio.

To determine the percentage humidity in the previous example of air at 0°F, 100% RH when heated to 68°, the values can be attained from ASHRAE tables or from psychrometric charts that go as low as zero. (Also refer to the ensuing sections on psychrometric charts.)

Pound/Grain Conversion

According to the ASHRAE psychrometric tables at 0% and 100% RH, a pound of dry air can hold 0.0007872 pounds of vapor. The decimal fraction the tables yield is rather awkward to work with so a more workable number can be derived by converting the decimal fraction to grains. One grain is approximately equal to one drop of water because one pound of water equals 7,000 grains. Because the decimal fractions are difficult to work with, many psychrometric charts and tables make the conversion and have values expressed in grains instead of pounds.

Percentage Humidity Example

In the previous section, the pound/grain conversion can be shown mathematically as:

$$0.0007872 \times 7,000 = 5.51 \text{ grains}$$

Fundamentals

By making the conversion, a more manageable whole number results which can be used for mathematical calculations. At 68°F a pound of dry air can hold 0.01475 pounds of vapor. Convert this quantity to grains,

$$0.01475 \times 7000 = 103.25 \text{ grains}$$

The converted number can then be used to determine the percentage humidity (at 68°F):

$$5.51 \div 103.25 = 5.34\%$$

This value differs slightly from the relative humidity for the same conditions of 4.8%. As a result, no great error results if the values are used interchangeably. Only in scientific or highly technical applications would the difference be considered a significant factor.

SECTION 2

PSYCHROMETRIC CHARTS

Air consists of two principle elements — dry air and water vapor. These gases obey Dalton's law and act independently of each other.

Water vapor is a small part of the total air mass, but it is always present in the atmosphere. It can be less than 1% of the weight of atmospheric air in temperate climates. However, it is always less than 3% — even in extreme climatic conditions. Nevertheless, it is the most variable factor in air, and these variations create climate conditions which significantly affect human comfort and the physical characteristics of many substances.

The study of moisture content and measurement in air is called psychrometry. The science which concerns itself with the thermal properties of moist air, its effect on human comfort and materials, and its measurement and control is called psychrometrics. No one can be an expert in air conditioning or environmental control without a fundamental knowledge of this science. The air conditioning engineer's most important tool in this area is the psychrometric chart which is a graphic representation of all values required to solve heating, cooling and air conditioning problems. The advantage of psychrometric charts — they provide answers to environmental control problems without relying on the use of complex equations and mathematical computations. If an application engineer or technician has two known values, these values can be plotted on a psychrometric chart to determine other unknown values.

Psychrometric charts are triangular or boot shaped, with the instep line or curve representing 100% relative humidity.

Psychrometrics

However, the exact configuration depends on the temperature ranges and the values required. At first glance the chart seems a maze of criss-crossing lines which are the horizontal, vertical and diagonal coordinates related to key psychrometric values. Upon closer scrutiny, these intersecting air property lines are transformed into an invaluable design and application tool. In the hands of a knowledgeable user, this tool can yield air values more rapidly than can a computer.

There are approximately nine values depicted or charted on a psychrometric chart. Although each value will be addressed briefly, this text will concentrate primarily on the values which specifically relate to humidification. The following is a list of air properties typically depicted on most psychrometric charts:

Value	Abbreviation
Dry Bulb Temperature	DB
Wet Bulb Temperature	WB
Dew Point Temperature	DP
Specific Humidity Ratio	SPH
Relative Humidity	RH
Specific Volume (Air Density)	V
Heat Content (Enthalpy)	h
Vapor Pressure	PW
Sensible Heat Ratio	SHR

Dry Bulb Temperature

The base line on the psychrometric chart indicates dry bulb temperature. It is appropriately the most fundamental value in the chart. This value is the temperature which a room thermometer senses. It is strictly a measurement of heat intensity, and it in no way relates to vapor in air or in latent or total heat quantity. For example, a lighted match may have a much higher temperature than a hot water boiler. However, a match cannot heat a room while a boiler supplies sufficient heat quantity to heat a building.

Humidity Control

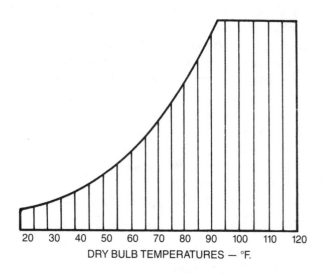

Figure 2-1. Location of dry bulb (DB) temperature lines on a psychrometric chart.

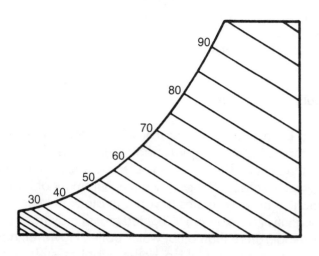

Figure 2-2. Location of constant wet bulb (WB) temperature lines on a psychrometric chart.

Psychrometrics

The dry bulb temperature lines extend vertically from the base line to the top of the chart as shown in Figure 2-1. These are constant dry bulb temperature lines. Further, any and all points on a given line yield the same temperature as the base line reading. The symbol for dry bulb is DB, and the Fahrenheit scale is used most prevalently with this value.

Wet Bulb Temperature

The wet bulb temperature scale lies along the outer curved line towards the left margin of the chart. Figure 2-2 illustrates the constant wet bulb temperature lines which extend downward and to the right at approximately a 30° angle from horizontal. Every point on a given wet bulb line is at the same wet bulb temperature. The 60° wet bulb line, for example, is an indicator of 60° whether a point occurs at the top, bottom or anywhere in between those two points. The symbol for wet bulb is WB.

Psychrometers

The standard wet bulb temperature is determined by covering the bulb of an ordinary thermometer and exposing it to a stream of air with a velocity of 1,000 ft/minute. The vaporization rate and cooling effect on the thermometer relates directly to the amount of moisture in the surrounding air.

In humidification the wet bulb reading has its greatest significance when compared to the dry bulb reading. Consequently, one of the most valuable service and application tools is the psychrometer or hygrometer. In this instrument two thermometers are combined side-by-side on a mounting frame. One thermometer has the dry bulb and measures the actual air temperature; the other has wetted wicking attached to its bulb and its reading is influenced by the water vapor in the air. With the advent of solid-state technology in recent years, however, highly accurate electronic meters are available which provide these values on digital displays. An example of an electronic humidity meter is shown in Figure 2-3.

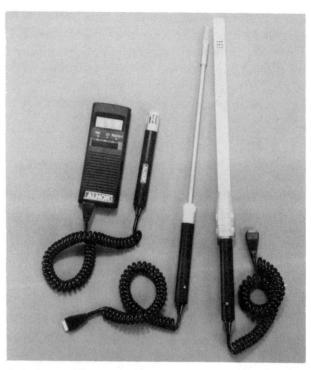

Figure 2-3. Example of digital humidity meters. (Courtesy, Alnor Instrument Co.)

A sling psychrometer is illustrated in Figure 2-4. The sling psychrometer is a very popular service instrument because it can quickly obtain accurate readings in any job situation. Physically it consists of two matched thermometers mounted on a common frame which swivels from the end of a handle. The bulb of one thermometer is covered with a wetted wick, and the unit is whirled through the air. The whirling action is equivalent to exposing the thermometer bulbs to a 1,000 ft/min air stream. With total saturation (100% relative humidity) the air cannot accept vapor from the wetted wick. Consequently no evaporation or cooling takes place, and the resulting dry bulb and wet bulb temperatures are the same. However, if the air is not saturated with vapor, moisture will evaporate from the wetted wick. This evaporative action cools the wet bulb and temperature reading falls below that of the dry bulb temperature.

Psychrometrics

Figure 2-4. Typical sling psychrometer.

Wet Bulb Depression

The temperature difference between the wet and dry bulb temperatures is referred to as wet bulb depression. The lower the vapor content in air, the greater the wet bulb depression. Subsequently, the wet bulb temperature related to dry bulb temperature is a precise indicator of humidity. This value is tabulated in Table 2-1. The intersection of the vertical dry bulb temperature line and the horizontal wet bulb depression line indicates the relative humidity (RH) value. This value can be determined even more quickly with a psychrometric chart by reading the RH value at the intersection of the dry bulb and wet bulb temperature lines.

In heating and air conditioning applications, the wet bulb temperature is the most critical value because it is a direct measure of total heat present in the air. This assertion holds true because the wet bulb is proportionately affected by dry bulb temperature (sensible heat) and vapor content (latent heat).

Humidity Control

Table 2-1. Chart used to determine relative humidity with known dry and wet bulb values.

Wet bulb depression deg.	Dry bulb temperature, F										
	68	69	70	71	72	73	74	75	76	77	78
1	95	95	95	95	95	95	95	96	96	96	96
2	90	90	90	90	91	91	91	91	91	91	91
3	85	85	86	86	86	86	86	86	87	87	87
4	80	81	81	81	82	82	82	82	82	83	83
5	76	76	77	77	77	78	78	78	78	79	79
6	71	72	72	72	73	73	74	74	74	74	75
7	67	67	68	68	69	69	69	70	70	71	71
8	62	63	64	64	65	65	65	66	66	67	67
9	58	59	59	60	61	61	61	62	62	63	63
10	54	55	55	56	57	57	58	58	59	59	60
11	50	51	51	52	53	53	54	54	55	56	56
12	46	47	48	48	**49**	50	50	51	51	52	53
13	42	43	44	45	45	46	47	47	48	48	49
14	38	39	40	41	42	42	43	44	44	45	46
15	34	35	36	37	38	39	39	40	41	42	43
16	31	32	33	33	34	35	36	37	38	39	39
17	27	28	29	30	31	32	33	34	34	35	36
18	23	24	25	27	28	29	29	30	31	32	33
19	20	21	22	23	24	25	26	27	28	29	30
20	16	18	19	20	21	22	23	24	25	26	27
21	13	14	15	17	18	19	20	21	22	23	24
22	10	11	12	13	15	16	17	18	19	20	21
23	6	8	9	10	12	13	14	15	16	17	18
24	3	5	6	7	9	10	11	12	13	14	16

Dew Point Temperature

When air is cooled to the point of complete saturation, it has 100% relative humidity. In other words, the air has all the water vapor it can hold. Any further cooling would cause the vapor to release its latent heat and condense into liquid. The specific temperature where condensation begins is the dew point.

Since dew point indicates the temperature at which air contains all the vapor it can hold, it is a true measure of the latent heat in air. Moreover, at 100% saturation the dry bulb temperature, wet bulb temperature and dew point are one and the same temperature.

Psychrometrics

A dew point indicator is an instrument used to determine the dew point. Typically, the instrument has a shiny metal surface which is gradually cooled until it starts to cloud with condensation (i.e., reaches the dew point). Consequently, dew point can be defined as the temperature at which water condensation begins to form when air is cooled.

Abbreviated as DP, dew point temperature has a special significance in humidification. In cold weather, if window surfaces or inner walls are below the dew point, an annoying, damaging condensation condition exists. In that light, the limiting factor of humidification hinges upon the ability of walls and windows to prevent or resist condensation. Concealed condensation within walls promotes mold growth and causes serious moisture damage. Window condensation, on the other hand, is easily perceptible and can serve as a warning of potential construction deterioration.

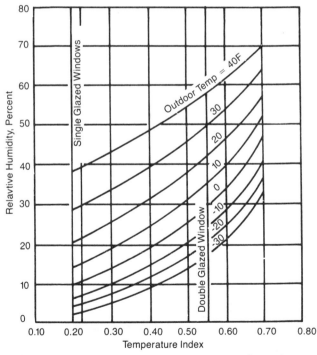

Figure 2-5. Chart showing the intersection of inside temperatures and relative humidities.

Humidity Control

Figure 2-5 charts the intersections of inside temperatures and relative humidities where condensation tends to form on various types of single- and double-glazed windows. Outside walls with inadequate or improperly installed vapor barriers are especially vulnerable to condensation damage within the walls. The application engineer should rightly be responsible for analyzing window and wall characteristics in order to ascertain the maximum humidities permissible with low outside air temperatures.

Figure 2-6 shows the dew point temperature part of a psychrometric chart. The dew point temperature lines originate on the left at the outer curved line. At this point along the curve, the dew point and wet bulb markings are the same. Where they differ, however, is the dew point lines run horizontally to the right while the wet bulb lines run diagonally and downward to the right approximately at a 30° angle. Any point on the horizontal constant dew point line corresponds to the dew point temperature on the left hand scale. On some charts the dew point scale is also indicated on the right hand side.

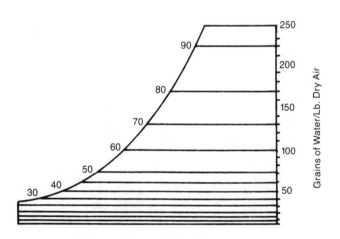

Figure 2-6. Dew point temperature lines (dew point and wet bulb are the same along the curved line).

Psychrometrics

Specific Humidity (Humidity Ratio)

On the right hand side of the chart in Figure 2-6 is the specific humidity scale. The specific humidity scale indicates the amount of moisture in one pound of dry air. This can be expressed in pounds of moisture or in grains depending on the preference of the chartmaker. Specific humidity is a very practical value in calculating equipment capacity. When cooling or humidifying an engineer must determine how many pounds of moisture must be removed or added to an environment in order to meet the design requirements.

In a chart, the constant specific humidity lines are horizontal and they coincide with the constant dew point lines because the dew point temperature represents complete saturation of the air. The moisture per pound of dry air is abbreviated as W.

Relative Humidity

As depicted in Figure 2-7, the constant relative humidity line in a psychrometric chart sweeps upward and to the right in a smooth curve. Relative humidity can be determined quickly when two known values are plotted on the chart. For example, take 78°DB temperature and 65°WB. If the DB temperature line is projected upward and the WB line is projected downward towards the right, the vectors intersect almost exactly on the 50% relative humidity (RH) line.

If the intersecting lines of two known values fall between the RH curves, the value must be determined by the process of interpolation. This entails estimating the intermediate value between the two known points. For example, suppose an intersection falls halfway between the 20% and 30% RH curves. The total difference between these two values is equal to 10%. Half of this difference (5%) is added to the lower value (20%) resulting in an RH of 25% at the mid-point intersection. If the intersection is ¾ the distance, the interpolated value is ¾ of 10% which is then added to the 20% and equals 27.5% RH.

Humidity Control

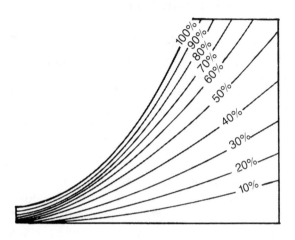

Figure 2-7. Constant relativity humidity lines.

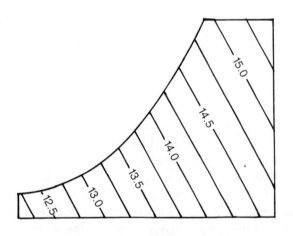

Figure 2-8. Specific volume lines (60° from horizontal).

Psychrometrics

The outer curved line which gives the psychrometric chart its triangular or boot-shape represents a saturation condition (100% RH). At this point the wet bulb and dew point temperatures are the same. These constant relative humidity lines decrease in value by 10% increments as they move downward and away from the saturation line.

Specific Volume Lines

Specific volume lines are represented in the chart shown in Figure 2-8. These constant specific volume lines occur at an angle approximately 60° from horizontal. These vectors express the cubic feet (ft^3) of a mixture per pound of dry air. Density or lbs/ft^3 of the mixture is the reciprocal of the values shown in the chart. If a density value is required, it can easily be calculated by dividing the specific volume shown in the chart into 1.

The specific volume lines increase in value from left to right. The values for these lines are generally expressed in 0.5 ft^3/lb but can vary among different chartmakers. As temperature increases, air expands. For example, at 56° the specific volume is 13.0 ft^3/lb of dry air; at 76° it increases to 13.5 ft^3. Conversely, the density decreases from .077 to .074 lbs/ft^3.

Enthalpy

Enthalpy is a measurement of heat energy expressed in Btu/lb of dry air. As Figure 2-9 illustrates, the constant enthalpy lines run from the left of the chart and downward at an angle of approximately 30° from horizontal. In actuality they are extensions of the wet bulb line because the wet bulb indicates total heat in air. The enthalpy scales can be on the extreme left on some charts or at the bottom right on other charts. The lines are typically scaled in 5 Btu increments. At 58°WB the total heat is 25 Btu/lb, and at 65°WB the total heat is 30 Btu/lb of air.

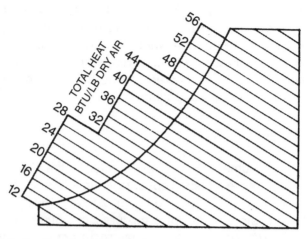

Figure 2-9. Constant enthalpy lines (positioned at approximatly 30° from horizontal).

Vapor Pressure

Vapor pressure (PW) is the specific pressure exerted by water vapor in a sample of air. This pressure is the force which promotes vapor migration from an area of high vapor content to an area of low vapor content. Saturated air at 60° has a vapor pressure of 0.522" Hg (mercury); saturated air at 0° only has 0.038" Hg vapor pressure. The difference, 0.484" Hg is significant in driving vapor molecules through any opening and into an area with lower vapor pressure. In fact, vapor can move toward a low vapor pressure area even though wind pressure or a moving air stream may oppose this movement.

Vapor pressure is directly related to dew point (point where water vapor condenses due to decreasing temperature). The dew point is also the point at which water turns into vapor due to an increase in temperature. Dew point, therefore, is analogous to the boiling point. If vapor pressure is reduced, for instance, water vaporizes or boils at a lower temperature. If vapor pressure is increased, a higher water temperature is required to vaporize (boil) the water.

Sensible Heat Ratio

Table 2-2 shows the vapor pressure expressed in inches of mercury at the extreme left margin of the chart. Although vapor pressure is an important factor, it is normally not required to solve most practical problems of humidification or air conditioning. Since vapor pressure is often not significant, it is omitted from many charts. Also note in some charts the extreme left margin may contain the designations for sensible heat ratio as shown in Table 2-2. This value indicates the ratio of sensible heat to total heat in a process. The symbol for sensible heat ratio is SHR.

Sensible heat ratio plays an important role in the design of air conditioning systems. Good design demands that the system remove the calculated amounts of sensible and latent heat simultaneously. The SHR is determined by dividing sensible heat gain by total heat gain. Then by projecting this value on the psychrometric chart, the combination of dry bulb and wet bulb temperatures required in the supply air to attain the proper SHR is shown.

Although most psychrometric charts are based on standard barometric conditions of 29.92" Hg, complete accuracy only results when the actual barometric pressure is at this level. However, for ordinary pressure deviations, variations from the chart are often insignificant. As a result, corrections are only necessary when working at extremely high altitudes.

The importance of the chart is quick reference to values required for solving practical humidification problems.

Figure 2-10 is a complete chart for the "normal" temperature range. If required, other charts are available for the low temperature range. Note that vapor pressure is not included in the chart. If the vapor pressure values are required for reference, they are published in the ASHRAE Handbook of Fundamentals in tables which are similar to Table 2-2 or in psychrometric charts which include vapor pressure values.

Humidity Control
Table 2-2. Psychrometric chart based on standard barometric conditions of 29.92" Hg.

(Reprint by permission—ASHRAE Handbook and Product Directory)

Fahr. Temp. t(F)	Absolute Pressure pa		Specific Volume, cu ft per lb			Enthalpy, Btu per lb			Entropy, Btu per (Lb) (°F)			Fahr. Temp. t(F)
	Lb/Sq In.	In. Hg	Sat. Liquid vf	Evap. vfg	Sat. Vapor vg	Sat. Liquid hf	Evap. hfg	Sat. Vapor hg	Sat. Liquid sf	Evap. sfg	Sat. Vapor sg	
33	0.092227	0.18778	0.01602	3180.5	3180.5	1.01	1074.59	1075.60	0.00205	2.1811	2.1831	33
34	0.095999	0.19546	0.01602	3061.7	3061.7	2.01	1074.03	1076.04	0.00409	2.1755	2.1796	34
35	0.099908	0.20342	0.01602	2947.8	2947.8	3.02	1073.46	1076.48	0.00612	2.1700	2.1761	35
36	0.10396	0.21166	0.01602	2838.7	2838.7	4.02	1072.90	1076.92	0.00815	2.1644	2.1726	36
37	0.10815	0.22020	0.01602	2734.1	2734.1	5.03	1072.33	1077.36	0.01018	2.1589	2.1691	37
38	0.11249	0.22904	0.01602	2633.8	2633.8	6.03	1071.77	1077.80	0.01220	2.1535	2.1657	38
39	0.11699	0.23819	0.01602	2537.6	2537.6	7.04	1071.20	1078.24	0.01422	2.1480	2.1622	39
40	0.12164	0.24767	0.01602	2445.4	2445.4	8.04	1070.64	1078.68	0.01623	2.1426	2.1588	40
41	0.12646	0.25748	0.01602	2356.9	2356.9	9.05	1070.06	1079.11	0.01824	2.1372	2.1554	41
42	0.13145	0.26763	0.01602	2272.0	2272.0	10.05	1069.50	1079.55	0.02024	2.1318	2.1520	42
43	0.13660	0.27813	0.01602	2190.5	2190.5	11.05	1068.94	1079.99	0.02224	2.1265	2.1487	43
44	0.14194	0.28899	0.01602	2112.3	2112.3	12.06	1068.37	1080.43	0.02423	2.1211	2.1453	44
45	0.14746	0.30023	0.01602	2037.3	2037.3	13.06	1067.81	1080.87	0.02622	2.1158	2.1420	45
46	0.15317	0.31185	0.01602	1965.2	1965.2	14.06	1067.24	1081.30	0.02820	2.1105	2.1387	46
47	0.15907	0.32387	0.01602	1896.0	1896.0	15.06	1066.68	1081.74	0.03018	2.1052	2.1354	47
48	0.16517	0.33629	0.01602	1829.5	1829.5	16.07	1066.11	1082.18	0.03216	2.0999	2.1321	48
49	0.17148	0.34913	0.01602	1765.7	1765.7	17.07	1065.55	1082.62	0.03413	2.0947	2.1288	49
50	0.17799	0.36240	0.01602	1704.3	1704.3	18.07	1064.99	1083.06	0.03610	2.0895	2.1256	50

Psychrometrics

51	0.18473	0.37611	0.01602	1645.4	1645.4	19.07	1064.42	1083.49	0.03806	2.0842	2.1223	51
52	0.19169	0.39028	0.01602	1588.7	1588.7	20.07	1063.86	1083.93	0.04002	2.0791	2.1191	52
53	0.19888	0.40492	0.01603	1534.3	1534.3	21.07	1063.30	1084.37	0.04197	2.0739	2.1159	53
54	0.20630	0.42003	0.01603	1481.9	1481.9	22.08	1062.72	1084.80	0.04392	2.0688	2.1127	54
55	0.21397	0.43564	0.01603	1431.5	1431.5	23.08	1062.16	1085.24	0.01587	2.0637	2.1096	55
56	0.22188	0.45176	0.01603	1383.1	1383.1	24.08	1061.60	1085.68	0.04781	2.0586	2.1064	56
57	0.23006	0.46840	0.01603	1336.5	1336.5	25.08	1061.04	1086.12	0.04975	2.0535	2.1033	57
58	0.23849	0.48558	0.01603	1291.7	1291.7	26.08	1060.47	1086.55	0.05168	2.0485	2.1002	58
59	0.24720	0.50330	0.01603	1248.6	1248.6	27.08	1059.91	1086.99	0.05361	2.0434	2.0970	59
60	0.25618	0.52160	0.01603	1207.1	1207.1	28.08	1059.34	1087.42	0.05553	2.0385	2.0940	60
61	0.26545	0.54047	0.01604	1167.2	1167.2	29.08	1058.78	1087.86	0.05746	2.0334	2.0909	61
62	0.27502	0.55994	0.01604	1128.7	1128.7	30.08	1058.22	1088.30	0.05937	2.0284	2.0878	62
63	0.28488	0.58002	0.01604	1091.7	1091.7	31.08	1057.65	1088.73	0.06129	2.0235	2.0848	63
64	0.29505	0.60073	0.01604	1056.1	1056.1	32.08	1057.09	1089.17	0.06320	2.0186	2.0818	64
65	0.30554	0.62209	0.01604	1021.7	1021.7	33.08	1056.52	1089.60	0.06510	2.0136	2.0787	65
66	0.31636	0.64411	0.01604	988.63	988.65	34.07	1055.97	1090.04	0.06700	2.0087	2.0757	66
67	0.32750	0.66681	0.01605	956.76	956.78	35.07	1055.40	1090.47	0.06890	2.0039	2.0728	67
68	0.33900	0.69021	0.01605	926.06	926.08	36.07	1054.84	1090.91	0.07080	1.9990	2.0698	68
69	0.35084	0.71432	0.01605	896.47	896.49	37.07	1054.27	1091.34	0.07269	1.9941	2.0668	69
70	0.36304	0.73916	0.01605	867.95	867.97	38.07	1053.71	1091.78	0.07458	1.9893	2.0639	70
71	0.37561	0.76476	0.01605	840.45	840.47	39.07	1053.14	1092.21	0.07646	1.9845	2.0610	71

* Compiled by John A. Goff and S. Gratch.
** Extrapolated to represent metastable equilibrium with undercooled liquid.

Humidity Control

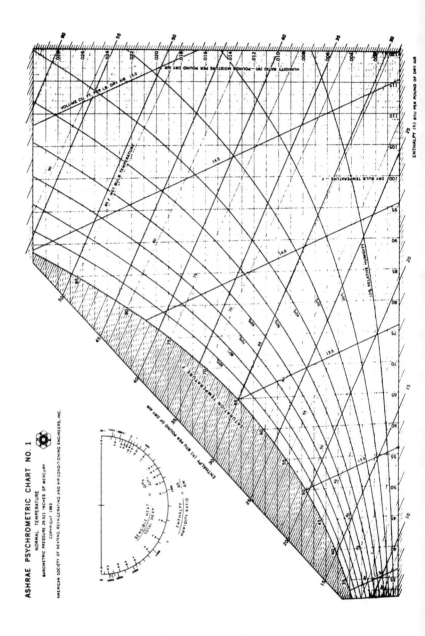

Figure 2-10. Psychrometric chart complete with all values.

PSYCHROMETRIC CHART CONSTRUCTION

The best way to develop a thorough understanding and a working knowledge of the "psych chart" is to construct one. Willis Carrier pioneered this procedure in 1911 when he published his Rational Psychrometric Formula. It merely involves graphically plotting psychrometric values on a chart. These values are readily available from thermodynamic moist air and steam tables. The basic values are humidity ratio, enthalpy, volume and vapor pressure. An excellent source for these tables is the chapter on psychrometrics found in the ASHRAE Fundamental Manual. Supplies necessary for this activity include: graph paper, a french curve and a straight-edge (ruler).

Figure 2-11 illustrates the basic steps in chart construction. The first step is to establish a horizontal dry bulb temperature base line. The next step is to establish a vertical axis which represents the humidity ratio (grains of moisture/lb of dry air when saturated). In Figure 2-11 the humidity ratio line is located on the left for ease of illustration, however, most charts depict humidity ratios on the right. This makes room for enthalpy values on the left side.

The next step—project a vertical dry bulb temperature line upward. The humidity ratio lines are then projected horizontally to the right. A point where the lines intersect establishes the 100% humidity or saturation condition for one pound of dry air at the indicated dry bulb temperature. The saturation line (X) is established when all the points are connected with a french curve giving the chart its "boot-like" configuration. The humidity ratio values along the vertical axis can then be reduced in 10% increments. The points where these lines intersect with those of the dry bulb temperature lines establishes the 90% to 10% saturation lines.

The 100% saturation line also represents the dew point line which is the same as the dry bulb temperature at this point. It projects horizontally across the chart. Cooling below the dew point causes moisture to condense. The saturation line also indicates the maximum heat content. Obtained from tables, these enthalpy values can be placed in the chart adjacent to the appropriate temperature settings.

The wet bulb temperature also originates at the dry bulb saturation line since the dew point and wet bulb temperatures are all the same at saturation. Although the wet bulb temperature vector begins at the saturation point, it projects downward and to the right at a slight angle because the wet bulb temperature is influenced directly by the latent heat contained in moisture, as well as by the dry bulb temperature of the air. This simply demonstrates that the wet bulb temperature indicates total heat in air at all times. Consequently, if moisture and latent heat are lowered below the saturation

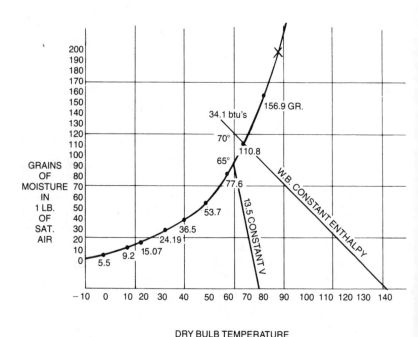

Figure 2-11. Basic values required for psychrometric chart construction.

Psychrometrics

value, the sensible heat must increase proportionately to maintain the same total heat content and wet bulb temperature. Lower moisture content causes a downward movement of the vector, and increased sensible heat promotes a movement towards the right. The result, wet bulb lines project downward from the saturation point at approximately a 30° angle. An example is shown in Figure 2-11. At 70°WB, the total enthalpy value at saturation is 34.1 Btu. With no moisture content, dry bulb temperature would have to be raised to approximately 142° to maintain heat content at 34.1 Btu. Extending the wet bulb line from 70° at saturation to 142° on the base line means that any point on the diagonal line is at 34.1 total enthalpy and at 70° wet bulb.

The constant volume or ft^3/lb of dry air can be established in a procedure similar to the wet bulb line. As shown in Figure 2-11, a volume for 65° air at saturation is determined to be 13.5^3. If air has no moisture content, air at 76° has 13.5^3/lb volume. By projecting a line from the 76° base to the 65° saturation point, a constant line at 13.5 is established on the chart. In this same manner constant volume lines in 0.5 cubic foot increments can be established.

After all the lines and temperature points have been determined, enthalpy values can be included to the left of the saturation line. The humidity ratio (either in grains, pounds or both) can be place to the right of the highest dry bulb temperature vertical line. If necessary, vapor pressure values can also be plotted on this right side. Sensible heat ratio indicators can also be placed on the right side. A reference dot is generally placed at the 80° and 50% relative humidity intersection. If only sensible heat is removed from the air, it is indicated by a horizontal line originating from the reference dot. If only latent heat is removed, it is indicated by a straight line emanating down from the reference dot. In actual practice both sensible and latent heat are removed in cooling applications, and these are indicated by a sloping line. If a designer needed to maintain a prescribed sensible heat ratio, design values can be plotted to depict the desirable slope on the psychrometric chart.

Humidity Control
PSYCHROMETRIC CHART APPLICATION

The tremendous advantage in using a psychrometric chart is its ability to provide instant analysis of existing air handling conditions. In addition, psychrometric charts provide instant reference for design condition requirements. Complete orientation, for example, is attained by plotting two known values. In Figure 2-12 two known values, 85°DB temperature and 50% relative humidity have been plotted, and the point where the lines intersect pinpoints the precise location on the chart. By reading to the left from this point, the following values can be determined: dew point is 64.5°, WB temperature is approximately 71° and enthalpy is 34.9 Btu. Reading to the right from the intersect reveals a humidity ratio of 0.0132 lbs or $0.0132 \times 7,000 = 92.4$ gr/lb of dry air. If the chart listed a vapor pressure table, it would show a vapor pressure of 0.61" Hg. Specific volume is shown as 14.0 ft^3/lb of air. Density can be instantly calculated by taking the reciprocal of 14.0 ft^3 which is 0.0714 lbs of air/ft^3. Therefore, by plotting only two psychrometric values, eight other values are quickly accessible.

On psych charts, changing conditions should serve as signals that sensible heating creates horizontal movement to the right and sensible cooling horizontally to the left. During humidification, adding moisture or latent heat causes vertical movement on a chart as the arrow indicates in Figure 2-12. Dehumidification, on the other hand, creates the opposite effect and induces a downward movement on a chart.

Changing air conditions, however, may involve simultaneous changes in sensible heat and moisture content. The new condition tends to move horizontally and vertically in proportion to the sensible and latent heat changes involved. Simply, the changing conditions generate a diagonal movement away from the initial point.

Psychrometrics

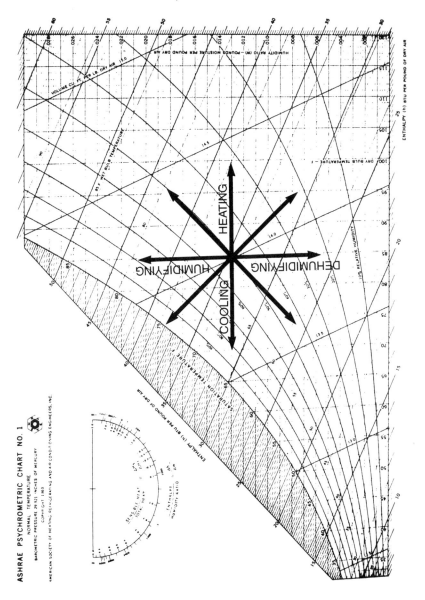

Figure 2-12. Psychrometric chart with center point located at 85°DB and 50% RH.

AIR MIXTURE VALUES

Calculation & Projection

Air handling applications, particularly in commercial and institutional situations, usually involve air mixtures. Recirculated and outside air are typically mixed in central fan systems in order to meet ventilation requirements at the most energy efficient level. A psychrometric chart analysis provides the new properties of the air mixture. The use of a psychrometric chart is advantageous in that it eliminates the time consuming, "hands-on" method of gathering actual instrumented measurements.

A central fan system with 10,000 cfm capacity serves as a typical example. If ventilation requirements involve 10% outside air, what are the properties of the mixture. Moreover, if recirculated air is 75°DB and 57°WB while outside air is 40°DB and 30°WB, how much heat and humidity is required to maintain comfort. In Figure 2-13 point A indicates recirculated air conditions while point B pinpoints outside air. The conditions of the mixture are represented somewhere along the straight line connecting points A and B.

The condition of the mixture can be determined in two ways. The new dry bulb temperature can be mathematically calculated and then plotted on the straight line or a vector analysis of the effect of outside air can be made. In the mathematical calculation, outside air has 10% effect and recirculated air 90%. Consequently, the procedure requires that 10% of the outdoor air dry bulb temperature and 90% of the recirculated air dry bulb temperature be added together. The result of the computation represents the dry bulb temperature for the mixture.

$$\begin{array}{ll} \text{Recirculated air DB} & 75 \times 0.90 = 67.5 \\ \text{Outside air DB} & 40 \times 0.10 = \underline{04.0} \\ \text{Mixture DB temp} & = 71.5 \end{array}$$

Psychrometrics

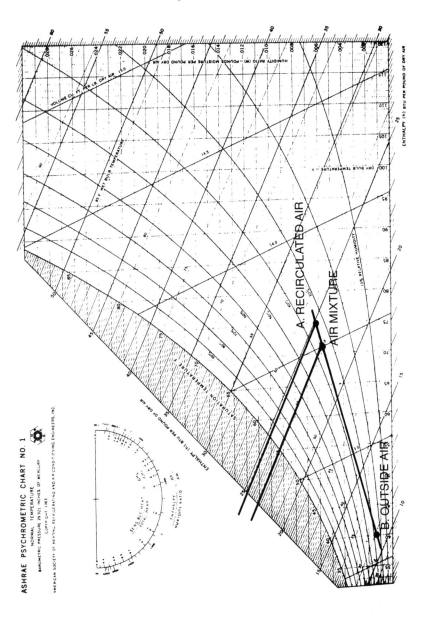

Figure 2-13. Psychrometric chart where A is recirculated air conditions, and B outside air. Conditions of the air mixture will occur along the line A-B.

Humidity Control

By plotting this temperature on the straight line all the related values become readily accessible. This same mixture value can be determined by use of the vector method. This procedure involves measuring the distance from point A to point B. The effect of outside air lowers the recirculated air conditions by 10% of line AB. This also yields a new value for the mixture of 71.5°DB.

To determine the proper quantity of heat and moisture necessary for maximum comfort, subtract the actual enthalpy and grains of moisture of the mixture from enthalpy and grains of moisture at 75°DB and 57°WB.

Enthalpy—

$$\begin{array}{r} 75°DB, 57°WB = 24.5 \text{ Btu/lb dry air} \\ 71.5°DB, 55°WB = 23.5 \text{ Btu/lb dry air} \\ \hline \text{required heat} = 1.0 \text{ Btu/lb dry air} \end{array}$$

Moisture—

$$\begin{array}{r} 75°DB, 57°WB = 42 \text{ grains} \\ 71.5°DB, 55°WB = 38 \text{ grains} \\ \hline \text{required moisture} = 4 \text{ grains} \end{array}$$

The enthalpy and moisture calculations yield the amount of heat and moisture required/lb of dry air. The next step is to determine the number of pounds involved in circulating 10,000 cfm at 71.5°DB and 55°WB. This point on the chart reveals the specific volume of the mixture to be 13.5 ft³/lb. The number of pounds in 10,000 cfm = 10,000 ÷ 13.5 = 740.74 lbs. The same answer can be achieved by converting specific volume into density or lbs/ft³ of dry air. This is done by taking the reciprocal of 13.5: 1 ÷ 13.5 = 0.074074 lbs/ft³. This value multiplied by 10,000 cfm also yields 740.74 lbs of air circulated/minute by the central fan system.

The heat requirement/minute is determined by calculating the enthalpy/lb required to maintain design comfort conditions (estimated at 1 Btu) multiplied by the number of lbs of air circulated/minute (740.74) by the system. Then multiplying by 60 yields the Btuh.

$$1 \times 740.74 = 740.74 \text{ Btu/min} \times 60 = 44{,}444 \text{ Btuh}$$

Psychrometrics

The required moisture content is determined by the same method. For example, 4 grains of moisture is multiplied by the pounds of air/minute. The resulting value is converted to pounds by dividing by 7,000. The pounds are then converted into gpm by dividing by a factor of 8.3.

$$4 \text{ gr} \times 740.74 = 2961.88 \text{ gr/min}$$
$$2961.88 \div 7{,}000 = 0.423 \text{ lbs/min}$$
$$0.423 \div 8.3 = 0.05 \text{ gpm}$$

The psychrometric chart, consequently, is an invaluable tool in hvac/r property determination, performance analysis and design requirements. It enables the technician, sales engineer or system designer to have an instant and accurate perspective of any hvac/r application. Moreover, it provides a graphic documentation of system operation and performance which is easier to assimilate than a host of written reports. The secret of the psychrometric chart is to use it for documented decision making. Remember, regular use of the charts develops proficiency.

SECTION 3

RESIDENTIAL HUMIDIFICATION

Comfort

The human body produces heat through the process referred to as metabolism. Fueled by fats and carbohydrates, a "chemical furnace" exists within the body which continuously generates body heat. This process is simply an oxidation of food elements. Because of metabolism and a supersensitive control system, internal body temperature is maintained between the range of 97° to 100°, with an average norm of approximately 98.6°F.

Internal body temperature is properly maintained by controlled heat rejection to the surrounding atmosphere. This heat rejection can amount to 500 Btu/person/hour. However, it can vary from 330 to 1,450 Btuh depending on the individual's activity. In high occupancy buildings, such as auditoriums, the occupancy heat load can be one of the most significant factors in sizing air-conditioning equipment.

Body heat is rejected to ambient surroundings in a number of ways — conduction, convection, radiation and evaporation. When body temperature rises because of increased activity or high ambient temperatures, the most effective heat dissipater is evaporation. In fact, at a temperature of about 85°, practically all body heat rejection is accomplished by the process of evaporation. This is understandable because a significant amount of latent heat is required to vaporize even a small amount of perspiration.

A heating system in a building does not actually heat its occupants. Typically, a heating system is set to maintain a temperature range between 68° to 78°F. The occupants, on the other hand, maintain a body temperature of 98.6°F.

Remember, heat cannot flow from a low level to a higher level. Consequently, heat actually flows from the occupants to their surroundings. The heating system merely establishes a temperature high enough so an occupant's heat loss is comfortably balanced by internal heat generation.

In addition to the sensible heat content in the surrounding air, the humidity level is also important in preventing body heat loss. Dry air accelerates body cooling due to the process of evaporation. However, with proper humidity, the vapor pressure in the air retards the rate of evaporation and the resultant heat loss. This merely indicates that comfort can be obtained at a lower room temperature.

Health

Proper humidity has a significant affect on the respiratory system of the human body. The tissue lining the respiratory system is normally moist. Moisture lubricates the tissue and provides a protective barrier against germs and viruses. This protective barrier is diminished with low room humidity. In low humidity conditions, microscopic cracks tend to form in tissue membranes, and this provides potential access for harmful viruses and germs. Asthma and skin problems are particularly known to be aggravated by low humidity levels.

Furnishing Protection

A dry, unhumidified atmosphere affects more than just human health. Furniture and furnishings are weakened and damaged in improperly humidified homes. Wooden furniture in dry homes tends to absorb water. When it gives up the water, total volume of the wood decreases and the joints shrink and loosen. In the most extreme cases, furniture may actually fall apart. In addition, as hard wood floors in a home lose moisture, the integrity of the wood can be damaged and squeaky spots in the flooring may become apparent. Moreover, carpets are especially vulnerable to improper humidity levels. Wool or fiber carpets in particular contain a

Humidity Control

minute amount of moisture. As they lose moisture, the fibers become increasing more brittle and bits and pieces of the fiber may break-off and create air-born particles. In all actuality, the lack of humidification has the potential to deteriorate any substance that is hygroscopic in nature, including books, paintings, tape or film.

Static Electricity Prevention

Electrostatic charges are generated when materials of high electrical resistance are rubbed against each other. The result of this electron accumulation may even discharge a spark. This is especially evident when walking across carpeting in a low humidity area. Not only does the carpet fabric deteriorate (as mentioned previously), but a high level electrostatic charge is generated. Static electricity affects many common materials such as papers, fiber, fabrics and films and results in what is commonly referred to as static-cling. This static condition poses difficulties in many applications such as material handling. In some instances where flammable gases are used, there is an ever present danger of explosions due to static electricity.

Increasing relative humidity tends to prevent accumulation of static charges. In properly humidified environments, a thin film of moisture is deposited on the objects in a room. This film then acts as a conductor and grounds the electrostatic charge rendering it harmless. To a great extent, proper humidity level depends on the type of material or substance considered. Relative humidity of 45% or more usually eliminates or reduces electrostatic effects in most substances. However, wool and select synthetic materials may require still more humidity to minimize electrostatic effects. An interesting point is that merely adding a little humidity to an environment (i.e., 25% to 30% RH) may actually create a worse electrostatic problem than had no humidity been added at all.

Thermostat Setbacks

It is a fact that lower room temperatures can result in adequate comfort conditions — provided proper humidity levels are established. Lower room temperatures result in a lower temperature differential between inside and outside conditions. When this temperature difference is reduced, the heat loss in a home is also reduced. Heat loss is a function of heat transmission values and is referred to as the conductivity (K) factor. This represents the Btu transmission/hr/degree temperature difference through building material. Consequently, if the temperature difference is great, the heat loss is also great. If temperature differences can be minimized, less heat energy is required. As a result, dialing-down the thermostat results in energy conservation.

However, as heat is required to vaporize water (1,000 Btu/lb water), any energy conservation that results must be a net energy saving. In a well-constructed building which has adequate vapor barriers, weather-stripping, storm windows and doors, a net energy saving is possible. In other cases, though, a net energy loss can occur. This latter condition results when the building has a high exfiltration and infiltration rate caused by loose construction. Again proper weather-stripping and tight storm doors and windows can reduce the humidification load and translate into a net energy gain.

ASHRAE HOME

In the ASHRAE Handbook of Fundamentals a home located in the vicinity of Syracuse, New York was subjected to a design outside temperature of $-10°F$. The two-story home had an indoor temperature of 75°F. The home featured a brick veneer construction but had little or no insulation. The infiltration was estimated at 19,351 ft^3/hr. According to the chart in Table 3-1, each 1,000 ft^3 of infiltrating air at $-10°F$ and 80% RH contained 0.032 lbs of moisture. At a temperature of 68°F and 35% RH, 1,000 ft^3 of air requires 0.381 lbs of moisture. This indicates that 0.349 lbs (0.381−0.032) of moisture or vapor is required for each 1,000 ft^3 of infiltrating air if 35% RH is to be maintained at an indoor temperature of 68°F. The

Humidity Control

amount of moisture that must be added to the indoor environment to maintain a 35% RH:

$$19,351 \text{ ft}^3/\text{hr} \div 1,000 = 19.351$$
$$19.351 \times 0.349 = 6.75 \text{ lbs vapor}$$

The amount of heat required to vaporize this amount of water is determined as follows: If incoming water temperature is 40°F, each pound must be heated to the vaporization temperature of 140°F (100 Btu/lb is necessary) —

$$6.75 \text{ lbs} \times 100 \text{ Btu/lb} = 675 \text{ Btu}$$

Table 3-1. Total amount of vapor which 1,000 ft^3 can hold at various temperatures and relative humidities.

Moisture Content* % Relative Humidity

Temp. °F	20	25	30	35	40	45	50	55	60	70	80	90	100
−10	.008	.010	.012	.014	.016	.018	.020	.022	.024	.028	.032	.036	.040
0	.012	.015	.018	.021	.024	.027	.029	.032	.035	.041	.047	.052	.059
10	.020	.025	.029	.034	.039	.044	.049	.054	.059	.069	.078	.088	.098
20	.031	.039	.047	.055	.063	.071	.079	.087	.095	.110	.126	.142	.161
30	.051	.064	.077	.089	.103	.116	.128	.141	.154	.180	.203	.232	.258
40	.077	.097	.117	.135	.155	.174	.193	.213	.233	.272	.310	.358	.390
50	.123	.141	.171	.198	.227	.259	.284	.312	.341	.399	.456	.521	.574
60	.162	.205	.246	.288	.329	.370	.412	.453	.495	.579	.663	.747	.831
62	.176	.220	.264	.308	.353	.398	.442	.486	.532	.622	.711	.801	.893
64	.189	.236	.284	.332	.379	.427	.475	.524	.572	.668	.764	.863	.960
66	.203	.254	.306	.356	.407	.458	.510	.561	.613	.717	.821	.926	1.030
68	.217	.272	.326	.381	.436	.491	.547	.602	.657	.768	.880	.991	1.106
70	.232	.291	.349	.408	.467	.527	.585	.645	.704	.824	.944	1.063	1.186
72	.249	.312	.375	.438	.501	.564	.628	.691	.755	.884	1.013	1.139	1.273
74	.267	.334	.401	.469	.536	.605	.671	.740	.809	.947	1.084	1.224	1.364
76	.285	.357	.429	.502	.574	.647	.720	.792	.866	1.012	1.160	1.310	1.461
78	.305	.382	.458	.535	.614	.692	.769	.848	.925	1.082	1.241	1.400	1.564
80	.326	.408	.490	.574	.656	.740	.823	.905	.991	1.160	1.330	1.500	1.676
82	.348	.436	.524	.612	.701	.790	.881	.970	1.060	1.240	1.422	1.609	1.792
84	.371	.464	.558	.654	.749	.844	.939	1.033	1.129	1.322	1.520	1.710	1.916
86	.389	.496	.597	.698	.799	.901	1.003	1.105	1.212	1.418	1.627	1.833	2.048
88	.422	.523	.635	.743	.851	.960	1.068	1.178	1.288	1.510	1.731	1.958	2.189
90	.450	.563	.678	.793	.908	1.024	1.141	1.258	1.375	1.613	1.853	2.095	2.338

*Pounds per 1000 cubic feet of air (based on standard conditions)

Residential Humidification

At 140°, it takes about 1,015 Btu to vaporize each pound of water. For 6.75 lbs, 6,851 Btu are required to vaporize the water. The 675 Btu of sensible heat must also be added resulting in total of 7,526 Btu.

The ASHRAE home has 113,950 Btu heat loss when maintaining 75° indoor temperature at −10°F outdoor design temperature. This results in a temperature differential of 85°. The heat loss/degree is equivalent to 113,950 ÷ 85 = 1,341 Btu. Lowering 75° to 68° yields a 7° reduction. Reducing room temperature to 68° produces a savings equal to 7 × 1,341 Btu = 9,387 Btu. The net savings should then be equal to 9,387 Btu − 7,470 Btu expended in humidification. This is a net savings of 1,917 Btu, which is for all practical purposes very marginal.

If the infiltration rate were decreased by tighter construction and the use of vapor barriers, the savings might prove to be more significant. As a consequence, the advantage of good construction and humidification are two-fold—comfort and lower energy consumption.

In addition to fuel savings, other economies are involved. A healthier living environment and the preservation of equipment and furnishings are also factors which can be measured in dollars and cents.

HUMIDIFIERS

Today, a number of different kinds of humidifiers are available to humidify air. These include:
- pan type
- atomizer type
- wetted-element type
- infrared type

Pan Type Humidifier

A type of humidifier which uses evaporation is the pan type humidifier shown in Figure 3-1. The capacity of the pan type depends on its surface area as well as the temperature, humidity and velocity of air in the system. In a basic pan humidifier, a shallow pan is normally installed in the furnace plenum. A relatively simple flow control device, connected to the water supply, maintains a constant water level. Some pan types may have water absorbent plates (wafers) partially immersed in the water and partially exposed to the air stream. Through wicking (capillary) action, the upper portion of the wafer which is exposed to the air stream becomes wet. As air flows through the wafers, vaporization takes place. This humidifier design has very limited capacity and is only suitable for smaller homes that call for a very light humidification load. However, capacity can be increased by immersing an electric heater or steam coil in the pan to increase water temperature and the rate of evaporation. This combination of heat energy and water is effective — provided the water supply is relatively free of mineral contaminants. If high mineral content water is used, the high temperature element may promote scaling which can eventually burn-out the element.

Atomizing Humidifier

The atomizing humidifier primarily relies on water pressure to produce a fine mist or droplets which is then introduced directly into the air to be humidified. In some models, a

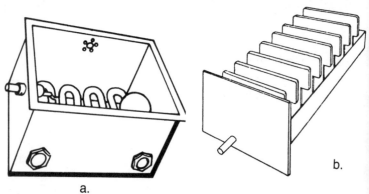

Figure 3-1. Typical configurations of pan type humidifiers: a) steam pan type equipped with electric heating element (steam coil), b) pan type with plates that are wetted by capillary action.

Residential Humidification

spinning disc or cone is used to create a fine mist (Figure 3-2). The ability of the air to absorb moisture depends on three factors:
- moisture content of entering air
- air temperature
- air velocity

A frequent problem with the atomization process is caused by water impurities. Impurities introduced into the air stream eventually settle out into the environment being humidified. An atomizing humidifier is a very effective, low cost device, if the water is relatively mineral free. However, with a high impurity content, the fall-out or dusting of lime and other contaminants can create a nuisance. Another disadvantage of the atomizer is that water particles which fail to vaporize tend to settle out in the air distribution system and can cause rust and corrosion problems.

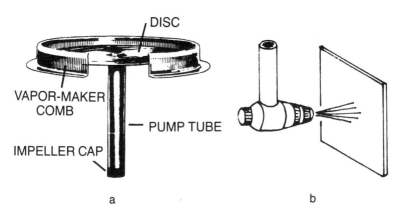

Figure 3-2. Some atomizing type humidifiers utilize a spinning disc (a) to cast water against a vapor making comb which atomizes water so it can be dispersed directly into the distribution system; b) high pressure nozzle sprays water through a pad or at a splash plate to atomize water.

Wetted-Element Humidifier

The wetted-element humidifier makes use of a porous, wetted media through which the air is circulated (Figure 3-3). The vaporizing surface of the wetted media may be a rotating disc positioned over a reservoir, a rotating paddle wheel, a drum or a belt rotating through a water reservoir. The element may also take the form of fixed pads which are wetted by sprays or by gravity-flow water distributed from a header or manifold.

Mounting positions of wetted-element humidifiers vary according to design. Some units mount on the underside of a supply duct or on the plenum itself. While other units have a fan or blower that draws air from the furnace plenum, through the pad and back to the plenum again. The by-pass type is also a common design. These usually have no moving parts, except for a water feeder or solenoid valve. They are mounted on the supply plenum of the furnace with an air connection to the return air plenum or duct. The difference in static pressure created by the furnace blower serves to circulate the warm air from the supply plenum through the wetted pad and to the return air duct. Humidified air is then routed back into the supply plenum and distributed to the areas to be humidified.

The under-duct model typically features a stainless steel, plastic or a fiberglass type reservoir mounted horizontally under the warm air supply duct. An electric motor with a speed of 2 to 5 rpm is located on the side of the reservoir. This motor turns a spit or shaft equipped with bronze or aluminum wire mesh discs that rotate in the reservoir. These discs are approximately 10" in diameter, and the bottom half always churns in the reservoir water. The square mesh openings pick up a thin film of water which is held in position by surface tension. As the discs turns, a constant supply of moisture is brought up above the water level and introduced to the hot air stream. Vaporizing capacity depends largely on the disc surface area as well as on the temperature, volume and relative humidity of air passing through the discs. This type of humidifier generates pure vapor void of any mineral content. The water

Residential Humidification

impurities are left behind in the reservoir or on the wetted elements.

In the rotating disc type, the mineral impurities build up on the disc material. In this type of humidifier, the mineral build-up works as an advantage because as mineral deposits build-up on the screen a greater surface for water retention is created slightly increasing capacity. Eventually, as more impurity content builds-up, the gap between the screens narrows and the capacity decreases. This necessitates that the elements in wetted-element humidifiers be cleaned or replaced periodically.

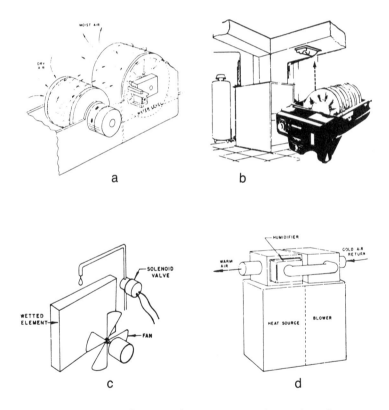

Figure 3-3. Various forms of wetting media utilized in common wetted-element humidifiers: a) wetted drum type, b) wire mesh discs, c) power type element, d) bypass type.

Infrared Humidifier

Infrared radiation is the main heat ingredient in solar energy. As such, it provides the energy required to vaporize water from rivers, lakes and ocean surfaces. Because of its ability to vaporize water, infrared radiation can be effectively employed in comfort humidification.

In the infrared humidifier, infrared lamps direct radiation onto a horizontal water reservoir. Vapor is not formed by boiling. Rather, vapor flashes off the water surface at a temperature that seldom exceeds 170°F.

Unlike most residential humidifiers, the infrared type does not depend on energy from the heating system to vaporize water. The infrared lamps provide their own heat for vaporization. This feature enables infrared humidifiers to be installed in low temperature locations. Figure 3-4 illustrates the installations of common infrared humidifiers. Since down-flow furnaces, electric furnaces, heat pumps and horizontal furnaces may have inherently low air temperature mounting locations, they are prime candidates for self-generating, compact infrared humidifiers.

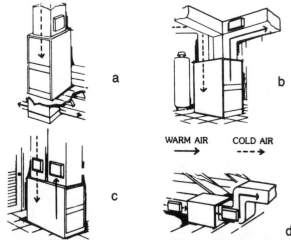

Figure 3-4. Common installations for infrared humidifiers: a) counter-flow installation on down-flow furnace, b) duct installation, c) standard installation for up-flow furnace, d) horizontal flow—typically used in attics or crawl spaces.

HUMIDIFIER SIZING

HUMIDIFICATION LOAD

Humidification load calculations are similar to those of a heat gain or loss survey. The humidification load, however, depends primarily on the quantity of outside air entering the conditioned space by way of mechanical ventilation or natural infiltration. This low temperature, low humidity air entering a space must be heated and supplied with moisture in order to meet design conditions.

Infiltration Rates

Residential buildings seldom have mechanical outside air supply dampers. Subsequently, any ventilation that occurs is attributable to natural infiltration. This infiltration varies with the temperature differential, wind velocity and building construction.

Infiltration rate can be estimated by several methods. The first is based on the measured leakage characteristics of building doors and windows. This procedure is known as the "crack method." Another method is known as the "air change method." This second method presumes a certain number of air changes/hour for each room or building, and depends on factors such as construction, outside exposure and usage.

Crack Infiltration Method

Since the major infiltration sources in a building are its windows and door cracks, actual measurement of these is considered the most accurate way of estimating air infiltration. This is especially true when window and door leakage values are properly determined from authoritative industry sources, such as the ASHRAE Handbook of Fundamentals.

Table 3-2 presents leakage values expressed in ft^3/hr/ft of window crack. These same values can be used for door cracks in residential applications. For a well-fitted door, the leakage factor is approximately the same as that of a poorly

Humidity Control

Table 3-2. Leakage values. (Reprint by permission—ASHRAE Handbook and Product Directory)

A. A 3/32" crack and clearance represent a poorly fitted window, much poorer than average.
B. The fit of the average double hung wooden window is 1/16" crack and 3/64" clearance as determined by measuring approximately 600 windows during the heating season.
C. The values given for frame leakage are per foot of sash perimeter, as determined for double hung wooden windows. Some of the frame leakage in masonry walls originates in the brick wall itself and cannot be prevented by caulking. Because caulking is rarely done perfectly and tends to deteriorate with time, it is considered advisable to choose the masonry frame leakage values for caulked frames as the average determined by the caulked and non-caulked tests.

Expressed in cubic feet per (hour) (foot of crack)

Type of Window	Pressure Difference (Inches of Water)				
	0.10	0.20	0.30	0.40	0.50
A. Wood Double-Hung Window (Locked) (Leakage expressed as cubic feet per hour per foot of sash crack; only leakage around sash and through frame given)					
1. Non-weatherstripped, Loose Fit[a]	77	122	150	194	225
2. Non-weatherstripped, Average Fit[b]	27	43	57	69	80
3. Weatherstripped, Loose Fit	28	44	58	70	81
4. Weatherstripped, Average Fit	14	23	30	36	42
B. Frame-Wall Leakage[c] (Leakage is that passing between the frame of a wood double-hung window and the wall)					
1. Around frame in masonry wall, not caulked	17	26	34	41	48
2. Around frame in masonry wall, caulked	3	5	6	7	8
3. Around frame in wood frame wall	13	21	29	35	42

Residential Humidification

fitted double hung window. However, if the door is poorly fitted, double this value for your calculations. If the door is properly weather-stripped, use ½ the value of a poorly fitted double hung window. If a single door is opened and closed with great frequency, such as in a high occupancy home or a small commercial establishment, the leakage value is three times that of a poorly fitted double hung window.

The values shown in Table 3-2 also depend on the pressure difference between inside and outside conditions. The first and last column (0.10) and (0.50) represent the extremes. In order to determine a representative value, a judgment call (based on temperature difference and prevailing wind conditions) is necessary. By multiplying window and door crackage by the representative value in Table 3-2, a total infiltration rate can be computed which is expressed in ft^3/hr. The total ft^3/hr is then used to calculate the humidification load. Table 3-1 provides the vapor content of 1,000 ft^3 of air at various relative humidity levels. If the specified low temperature of an area is 20°F, the outdoor air has an 80% RH and contains 0.126 lbs air/1,000 ft^3. If the temperature is 68°F and humidity is 35% RH, the moisture content must be 0.381 lbs/1,000 ft^3. Consequently, to meet indoor design specifications, the amount of vapor required for 20°F outside air is 0.381 lbs less the 0.126 lbs which already exists in the 20°F air sample.

$$0.381 - 0.126 = 0.255 \text{ lbs}$$

As a result, 0.255 lbs of vapor is required for each 1,000 ft^3 of infiltrating air.

If the infiltration rate of a building is 19,351 ft^3/hr, divide that number by a factor of 1,000 because the 0.255 lb value represents vapor required for each 1,000 ft^3 of in filtrating air.

$$19,351 \div 1,000 = 19.351 \times 0.255 \text{ lbs/hr}$$

The total humidification load is equivalent to 4.94 lbs/hr. Next, convert this into gallons per hour (gph):

$$4.94 \text{ lbs/hr} \div 8.345 = 0.59 \text{ gph}$$

Humidity Control

This load can easily be handled by most residential humidifiers.

Although the crack method of calculating the humidification load is regarded as the most accurate, building plans may not be readily available and actual job-site measurements are usually too costly and time consuming to perform. As a result, this method is not used commonly for a majority of applications.

Air Change Method

This method is reasonably accurate, and it is the quickest and most convenient approach to use especially when building plans and specifications are not easily accessible. This method is predicated on the fact that natural infiltration and exfiltration completely replace the cubical content of a room or building over a certain period of time. This replacement process is known as the air change rate, and it is based on the number of times the entire cubical content of the room or building is replaced over the duration of an hour.

Table 3-3, derived from the ASHRAE Handbook of Fundamentals, provides the air changes/hour for an individual room, hall or building. To be conservative in residential estimating, assume that total infiltration is the sum of the allowances for the individual rooms. This may not always be a valid assumption because infiltration on the windward side of the house or building may promote exfiltration on the

Table 3-3. Residential air changes that take place under average conditions, exclusive of air provided for ventilation. (Reprint by permission—ASHRAE Handbook and Product Directory)

Type of Room or Building	Air Changes/Hr.
Rooms with no windows or exterior doors	1/2
Rooms with windows or exterior doors on one side	1
Rooms with windows or exterior doors on two sides	1-1/2
Rooms with windows or exterior doors on three sides	2
Entrance halls	2

*For rooms with weatherstripped windows or with storm sash, use 2/0 these values.

Residential Humidification

leeward side. In this instance, actual infiltration may be less than the total of the room allowances. Moreover, it can be as little as half of the total infiltration allowance.

The most common method for estimating residential infiltration in the field is based on house construction. ASHRAE classifies homes into three categories and rates the hourly air changes as follows:

1. TIGHT HOUSE = ½ Air Change/Hr. This is a well insulated house which has: full vapor barriers, tight storm doors and windows, weather-stripping and a dampered fireplace.
2. AVERAGE HOUSE = 1½ Air Changes/Hr. The average house has some vapor barriers and insulation, nonweather-stripped storm doors and windows and a dampered fireplace.
3. LOOSE HOUSE = 2½ Air Changes/Hr. This is typical of pre-World War II homes which have no insulation or storm doors, no weather stripping or vapor barriers, and undampered fireplaces.

It is the sales engineer's responsibility to accurately analyze each building which is being humidified to determine its proper classification according to the preceding parameters.

SIZING FORMULA

If the air change rate has been determined, the following formula is used to size an application:

$$H = \frac{V \times (W_2 - W_1) \times R}{8.345}$$

Where:
H = required humidification rate (gph)
V = building's cubical content ($ft^3 \div 1,000$)
 to express $W_2 - W_1$, conversion to
 lbs/1,000 ft^3 is necessary
W_2 = lbs/1,000 ft^3 indoor design temperature and RH
W_1 = lbs/1,000 ft^3 outdoor design low temperature and RH
R = air changes/hr
8.345 = lbs/gal (to convert into gph)

Humidity Control

Calculating:
$$H = V \times (W_2 - W_1) \times R = \text{lbs vapor/hr}$$

Sizing Example

If a tight house with 3,600 ft², 8' ceilings, the value for V is equivalent to:
$$3600 \times 8 \div 1,000 = 28,800 \div 1,000 = 28.0$$

In order to maintain a 68°F and 35% RH, the value for W_2 (according to Table 3-1) is 0.381 lbs/1,000 ft³. If the outdoor design temperature is 20° and 80% RH, the value for W_1 is 0.126 lbs/1,000 ft³. Therefore $W_2 - W_1 = 0.381 - 0.126 = 0.255$ lbs/1,000 ft³. R = 0.5, so the total formula is:

$$\frac{28.8 \times 0.255 \times 0.5}{8.345} = 0.44 \text{ gph}$$

Square Foot/Cubic Foot Method

This is a simplification and often an over-simplification of the air change method. The square footage or cubical content of a building is all that is required to employ this sizing method.

Most residential humidifier manufacturers rate a unit according to its ability to humidify a given number of square feet or cubic feet of building area. This rating system is designed to help facilitate equipment selection. To make a selection, simply calculate the building area and chose a humidifier rated to handle that particular capacity.

Manufacturers typically develop area ratings based on the ability of a humidifying unit to vaporize given number of gallons/day. If a set of ratings for a unit comes without any specifications, the unit is probably capable of maintaining an acceptable humidity level in an average house of a given area. For example, if a unit is rated for 3,600 ft² (28,800 ft³), it would be oversized for a tight modern house and is approximately twice the size required. Conversely, it would be undersized by almost 50% in a loosely constructed building.

To present more accurate figures, some manufacturers qualify them for either tight, average or loose home construction. This helps to provide a greater degree of accuracy when making a sizing selection. The only factor neglected in this method of selection is the humidity variation encountered in different geographic areas. Generally the manufacturer's ratings are based on an outside design temperature of 20°F, 70% RH and an inside temperature of 72°F, 35% RH. An outside design temperature of 20°F is typical for the States of Louisiana, Georgia and Texas, but it does not reflect the lower temperatures in the Northwestern, Midwestern and Northeastern areas of the country where the greatest number of humidifiers are sold. Consequently, selections based on square footage ratings alone tend to ignore the important requirements for northern climates. This can be rationalized by the fact that 35% RH may be too high to maintain in extremely cold weather. For loose and average homes, a 35% RH is apt to cause excessive condensation on windows or within walls. If this is the case, 20% to 25% RH might be more practical. By merely using the manufacturer's selection procedure, correctly attaining this lower, more practical humidity range is possible. Nevertheless, this sort of constraint should not be placed on the sizing/selection process. The sizing and selection of a humidifier for any house should accurately reflect individual design conditions.

ASHRAE Method

Utilizing the same procedure as that outlined in the air change method, the ASHRAE method derives humidity values from a psychrometric chart. These values (humidity ratio) are expressed in pounds or grains of moisture/lb of dry air. A pound of dry air is considered the equivalent of 13.5 ft^3. This is actually the value for 70°F, but it is a fairly accurate value in the comfort range. The ASHRAE values read in quantity per 13.5 ft^3, while the air change method is based on moisture

Humidity Control

quantity per 1,000 ft³. Under the same conditions, both methods yield the same answer — provided the answers are converted to a common denominator such as gph or lbs/hr. The ASHRAE formula is:

$$H = \frac{VR(W_1 - W_0) - S + L}{13.51}$$

Where:
H = humidification load (lbs vapor/hr)
V = volume to be humidified (ft³)
R = infiltration rate (air changes/hr)
W_1 = humidity ratio (lbs moisture/lb dry air at indoor design conditions)
W_0 = humidity ratio (lbs moisture/lb dry air at outdoor design conditions)
S = moisture from internal sources (lbs moisture/hr)
L = other moisture losses (lbs/hr)
13.5 = ft³ dry air/lb

The last two values, S and L, represent variables more applicable to commercial-industrial applications than for residential. For example, S represents a significant source of internal moisture. This might come from adjoining structures whose high humidity infiltrates into the application area; it might stem from high occupancy applications wherein each occupant potentially contributes as much as 0.4 lbs/person/hr; or it might be due to a processing operation which releases moisture to the surrounding area. On the other hand, L represents other moisture losses which are possible in commercial and industrial buildings where certain hygroscopic products tend to absorb humidity. This condition is either negligible or nonexistent in most residential structures.

ASHRAE Method Example

Again considering a tight house with 3,600 ft² and 8' ceilings the volume is equal to 28,800 ft³. If indoor design is 68°F and 35% RH, the psychrometric chart reveals a W_1 value of

Residential Humidification

0.0052 lbs or 36.2 gr/lb of dry air. At outdoor design conditions of 20°F and 80% RH, the W_0 value is 0.00172 lbs or 12.04 gr/lb of dry air.

The R value, or air change/hour remains 0.5 or ½ the air change/hr. Assuming that a residential application has no significant S gains or L losses, the calculation for the house is:

$$H = \frac{28{,}800 \times 0.5 \times (0.0052 - 0.00172)}{13.4}$$

(Note: 13.4 is the most accurate ft^3/lb value for 68°F and 35% RH)

$$H = \frac{14{,}400 \times 0.00348}{13.4} = 3.74$$

$$H = \frac{3.74 \text{ lbs}}{8.345 \text{ lbs/gal}} = 0.45 \text{ gph}$$

This answer is practically the same as the previous example for the air change/hr method using the lbs moisture/ 1,000 ft^3 chart. In essence, then, the two methods are the same except that the moisture values are derived from different sources.

ARI Method

The ARI Method is based on capacity for a 24-hour day:

$$\text{gal/day} = \frac{24 \times V \times R \times G}{v \times 7{,}000 \times 8.34}$$

Where:
- 24 = hours/day
- V = total building volume (ft^3)
- R = air change/hr
- G = difference in moisture content between inside design and outside design conditions
- v = Specific volume of moist air at design conditions (13.4 ft^3/lb of dry air at 68°F, 35% RH)
- 7,000 = gr/lb
- 8.34 = lbs/gal of water

ARI Method Example

Considering a tight 28,800 ft³ house at indoor design conditions of 68°F, 35% RH and outdoor design of 20°F and 80% RH, the calculation for the house is:

$$\text{gal/day} = \frac{24 \times 28{,}800 \times 0.5 \times (36.2 - 12.04)}{13.4 \times 7{,}000 \times 8.34}$$

$$\frac{8{,}349{,}696}{782{,}292} = 10.67 \text{ gal/day} \quad \text{Then convert:} \quad \frac{10.67}{24} = 0.45 \text{ gph}$$

SUMMARY

For the most part, all methods of humidity load calculation yield the same results if they are equated correctly and accurately. However, minor variations between psychrometric chart values and the charted tables may cause calculations to vary to a slight degree. Nonetheless, each method is based on the fundamental premise that the total humidification load depends on how much outside air enters the building at low design temperatures. Whether expressed in ft³/minute or air changes/hour, the outside air must be supplemented with enough moisture to provide the required relative humidity at the design indoor temperature. This simply means that X number of grains or pounds of vapor must be added to each cubic foot of infiltrating air.

Although the method used depends primarily on an individual's preference, one of the most effective methods is the air change method which utilizes values from the lbs/1,000 ft³ chart. This chart provides quick reference to values ranging from −10° to 100°F at any RH. The psychrometric chart also provides a quick reference, but values below 30°F are typically not included on the normal temperature chart. These values must be selected from a low temperature chart. As a result, two charts must be consulted rather than one, and this is often not expedient for routine field applications.

SECTION 4

RESIDENTIAL HUMIDIFIER INSTALLATION & SERVICE

HEATING SYSTEM CONFIGURATIONS

The common types of heating/furnace configurations include:
- upflow furnaces
- downflow furnaces
- counterflow furnaces
- horizontal furnaces
- heat pumps
- hydronic units

The type of heating installation is obviously an important factor in determining just what type of humidification equipment can be added to a system.

Upflow Furnace Systems

The type of furnace is important in selecting humidification equipment. The conventional upflow furnace is probably the most adaptable because the plenum and the horizontal supply ducts lend themselves to many types of humidifier installations. This includes the by-pass type where air from the plenum is forced through the wetted pads and into the cold air supply duct or plenum due to static pressure. The humidified air then moves back through the air distribution system.

Downflow/Counterflow Furnace Systems

Both downflow and counterflow furnaces impose a limit on the use of a humidifier since heated air is discharged down through a bottom outlet. Consequently, there is no suitable mounting surface for the humidifier. Moreover, many downflow furnaces are located in tight quarters which further complicates humidifier installations. An infrared humidifier is best suited for counterflow furnaces because of the fact that it can be mounted on a cold air plenum. Although cold air does not have the vapor absorbing capacity of warm air, the infrared unit can generate vapor due to its independent energy source.

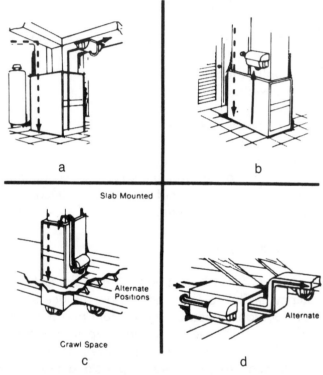

Figure 4-1. Typical furnace configurations showing humidifier installations: a) standard installation site on an upflow furnace (note, if several branch hot air ducts exist, mount unit under largest convenient duct and recirculated air will uniformly humidify home), b) plenum installation, c) installation in counterflow furnace—both in crawl space and for slab mounted, d) horizontal furnace humidifier installation.

Horizontal Furnace Systems

The horizontal furnace presents some of the same problems as does the counterflow type because it is usually located in tight quarters such as a crawl space or attic. However, the warm air discharge plenum and ducts usually provide sufficient accessibility for most humidifier installations.

Heat Pump & Electric Furnace Systems

A heat pump is primarily a refrigeration compressor in which the evaporator and condenser (heat rejector) operate in reverse by way of a reversing valve.

For summer cooling the inside coil functions as a cooling evaporator. The outside coil operates as a condenser for heat rejection. However, in a heating application, the coils are functionally reversed. The outside coil absorbs heat from the ambient air and rejects the heat (via the inside coil) to the air distribution system. The temperature of the air discharged from the condenser is usually too low for effective humidification with conventional humidifiers. A humidifier which provides its own heat energy (infrared and immersion elements for example) work most efficiently with heat pump systems.

Hydronic Systems

Hydronic units are either hot water or steam boilers which rely on a piping system to provide heat distribution. In the absence of an air distribution system, the only practical method for humidification is a self-contained unit. Any self-contained unit must have its own motor-blower and heat source. It can be placed in a closet or in the basement. Humidity distribution depends on the inherent tendency of water vapor to occupy the entire volume of an enclosure. One other readily available option is a room console unit. These require manual filling and have limited capacity for conventional buildings.

Humidity Control

HUMIDIFIER SELECTION & INSTALLATION

The humidifiers addressed in this section are central system humidifiers for residential or light commercial applications. Humidifiers for large commercial or industrial applications are addressed in an ensuing section.

Central System Humidifiers

This type of humidifier discharges moisture directly into the air stream of a distribution system and can be mounted in or on the following:
- supply air plenum
- supply air duct
- return air plenum
- return air duct

A central system humidifier may be installed as an integral part of a heating system or can be added on to the air distribution system as an ancillary unit.

EQUIPMENT SELECTION

The initial step in selecting equipment is accurately determining the total humidification load required for the installation. This task is accomplished by following the procedures presented in Section 3 — Sizing Humidifiers.

Although it may appear that equipment selection involves a simple match-up of load requirement to humidifier capacity, this is an area where professional judgment is often beneficial. Humidifier manufacturer's systematically test and rate equipment under steady state (consistent) conditions of air velocity, return duct dry bulb and wet bulb temperatures, and supply duct temperatures. The established Air Conditioning and Refrigeration Institute (ARI) standards used for determining the ratings are based on the following conditions: an air velocity of 800 ft/min, return duct temperature of 75°F DB and 56.5°F WB and the supply duct at 140°F DB. Figure 4-2 illustrates the recommended site for duct mounted humidifiers.

Installation & Service

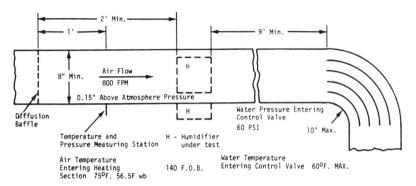

Figure 4-2. Recommended installation site for duct mounted humidifiers. (Reprint by permission — ARI)

The ARI also suggests test facilities and specifications for other humidifier types.

The ARI standards are an excellent vehicle for use in establishing honest performance ratings. In addition, the standards present a way to make objective comparison of humidifier capabilities. Remember, however, field operating conditions can vary considerably from the consistent conditions of the laboratory in which the tests where conducted. The 140°F supply duct temperature is especially significant as heat energy and the capability of air to hold moisture is greater at higher temperatures. If the supply air temperature averages significantly less than 140°F, then the ARI rating cannot be attained. The humidifier must then be derated to reflect the capacity which can be attained.

Knowledge of heating equipment characteristics is important when predicting the deviation from a rated capacity. A furnace may operate below 120°F discharge air temperature for a long period of time before a steady state condition creates a higher temperature. If the average furnace supply temperature is significantly below 140°F, then a humidifier rated at 140°F cannot develop its full rated capacity. If doubt exists, try to obtain operating characteristic information from the furnace manufacturer's application engineering department. Humidifier manufacturer's should be equipped to provide ratings based on lower supply temperatures.

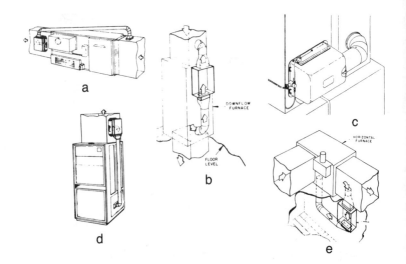

Figure 4-3. Humidifier installations showing bypass sites: a) horizontal furnace with bypass, b) bypass on a counterflow furnace, c) standard bypass on a low-boy furnace, d) bypass on a high-boy furnace, e) bypass on a roof-top furnace.

Another capacity factor to consider is that the humidifier only cycles when the system blower operates. If the blower cycles off, the humidifier should be inactive regardless of humidistat demands. If conditions such as wide differential thermostat operation or oversized heating equipment promote extended blower off-cycles, the humidifier is penalized in that it is not given sufficient operating time to produce its rated capacity. Only in the absence of such limiting factors should the humidifier's rating be taken at face value. However, if the face value rating conforms to an established industry standard, such as ARI Standard, it is an accurate basis for derating or up-rating humidifiers to reflect furnace performance variations.

After sizing procedure and evaluation has determined the system capacity in gph, or gal/24 hrs, the most suitable humidifier for the application must be selected (refer to Figure 4-4). In addition to capacity, other selection factors involve available mounting space, ease of installation, serviceability,

Installation & Service

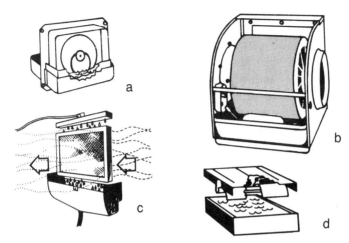

Figure 4-4. Common humidifier designs: a) atomizing, b) rotating drum (wetted element), c) fixed pad (wetted element), d) infrared.

costs, customer preference, ability to develop rated capacity, ability to give trouble-free operation and manufacturer's reputation.

On the basis of costs alone, the pan type, or the pan type with wafers or plates proves to be a good selection. Ease of installation and serviceability are also advantages offered with the pan type humidifier. However, its limited capacity (due to limited evaporating surfaces) is proven a deterrent because this design should only be considered for very light humidification loads.

The spray or atomizing humidifier offers many of the same advantages of the pan type including: low cost and ease of installation, service and maintenance. Unlike the pan type, atomizers are available with capacities adequate for any size application. The limitation of an atomizer is associated with high mineral content water. Not only is water sprayed into the air in small droplets or as a mist, but so are any impurities the water may contain. An annoying dusting and/or fallout of the impurities can result. In addition, droplets which do not completely vaporize settle out in the air distribution system and cause corrosion problems.

Humidity Control

The wetted-element humidifiers, which rely on moisture evaporation from screens or pads, are the most widely used humidifiers. They depend on the warm air flow through the wetted screen or pad for an effective evaporation rate. This design can rely entirely on the furnace blower to furnish the required air movement, or it can be equipped with an integral fan which draws air from the plenum, through the pad and back to the plenum to supplement the air movement produced by the blower. The screen type wetted-element unit has rotating aluminum or bronze discs. The elements are typically mounted on a shaft that is driven by a small, 3 rpm servomotor. The discs rotate through a reservoir (usually a pan) where the mesh of the screens pick up water. As the element rotates, the water filled mesh of the screens are exposed to the warm air flow which evaporates the water. A humidistat controls the motor operation.

The mounting location recommended for a wetted-element humidifier is under the horizontal supply duct—as near the furnace plenum as possible. It also can be mounted on the plenum, but this may require a plenum adapter. When installed in the plenum, it functions as a bypass type humidifier: warm air from the plenum is forced into the adapter hood, through the wetted screens or pad, then through piping into the cold air return and back through the heat exchanger to the distribution system.

Similar to the rotating screen element is the rotating drum (covered with a foam pad) which rotates through a water reservoir. Other rotating humidifiers feature paddle wheels or belts that rotate through the water reservoirs.

Space requirement must be carefully considered when considering the use of rotating type humidifiers. In the under-duct configuration, head room is sacrificed at the humidifier location. In the plenum mounts, the humidifier requires a substantial lateral area. In tight locations, this creates both installation and service problems.

The fixed pad bypass humidifier has the advantage of requiring less room laterally because it is basically a narrow pad that subsequently requires a narrow, compact housing. A header, fed by a humidistat controlled solenoid valve, allows water to trickle down by gravity through expanded metal pads. The static pressure difference between the supply plenum and the cold air return forces warm air through the wetted pad and back to the distribution system. One advantage of using this type of unit is that there is no motor and virtually no moving parts. The disadvantage, the header or water manifold must be carefully leveled for proper water distribution through the pad. Without adequate water distribution, the humidifier loses much of its humidifying capacity.

Another disadvantage to consider before selecting a bypass unit is that the effort of the furnace blower motor is hampered by the bypass installed between the warm air plenum and the cold air return. Even though this is not a major problem for most heating systems, it can pose a problem if the air bypass is not dampered off for summer cooling. The cooling coil requires a higher volume of air than that required for heating. If the bypass circuit is open, this can rob the cooling coil of sufficient air and heat load to keep its temperature above freezing. The result can be loss in cooling capacity and damage due to a frosted coil.

The location of a central air cooling coil should also be a primary consideration in locating all plenum mounted humidifiers. Do not position a humidifier so that the coil and coil mounting block air flow through a humidifier. Otherwise, the humidifier cannot operate efficiently because all wetted-element humidifiers require heat energy from the heating system to develop rated capacity. Applications exist where air does not have a high enough temperature to supply adequate heat energy. A heat pump, which uses reverse cycle refrigeration, is a case in point.

One solution is to consider using a pan type humidifier equipped with a self-contained electric heating element. This type is very effective since a major portion of the electrical energy it consumes is used for vaporization. The disadvantage, its high temperature heating element necessarily comes in direct contact with water and causes the minerals and impurities to fall out of suspension. The result, a scaling condition which can potentially lead to a burn-out of the electrical heating element. Connecting wetted-element humidifiers to a hot water source only compounds and accelerates a scaling condition.

Another alternative is the use of an infrared humidifier which radiates heat energy into water. Approximately 82% of its energy is effective in vaporizing water. The balance of its heat energy is introduced into the system as sensible heat. Although this sensible heat energy does not assist in the humidification process, it helps supplement the heating system. Consequently, the energy is not wasted. Like the humidifier that uses a heat element, an infrared unit is also adversely affected by water impurities. For instance, if a layer of scum forms over the water surface, it interferes with radiant heat transfer. This problem can be controlled to some extent, however, by equipping the unit with a flushing device to periodically purge the water reservoir and rid it of contaminates.

In summary, advantages and disadvantages of all units must be weighed and balanced during the selection process. Except for the atomizing humidifier, each design must be properly cleaned and maintained because the units are constantly subjected to water born contaminants. If not purged from the units, the concentration of contaminants can build up and pose serious problems. The use of demineralized water is a partial solution. However, it is not practical from a cost standpoint. Until fool proof, self-cleaning humidifiers evolve, humidifiers must be installed with serviceability in mind. Units must be readily removable for cleaning and routine maintenance at periodic intervals. The humidifiers must also meet national and local electrical and plumbing code requirements. Moreover, one prime consideration is a manufacturers' product warranty and responsiveness to information requests for installation and troubleshooting guidelines. In addition, thoroughly evaluate whether a particular type of humidifier might be suitable for the water conditions that prevail in the vicinity.

HUMIDIFIER INSTALLATION

The type of humidifier selected should properly depend on its adaptability to the existing heating system. Figure 4-5 illus-

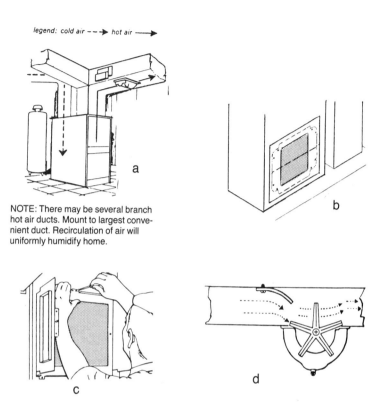

Figure 4-5. Humidifier installation: a) underduct site selection, b) many units come with a template to use as an installation guide properly sizing openings and locating mounting screws, c) some units require a mounting frame or support bracket, and some attach directly to plenum with screws, d) certain types may require a baffle to divert air to and through the unit (this illustration shows a sheet metal diverter fastened to the top of the duct).

trates a conventional heating system equipped with several supply ducts. Humidifiers should be installed in the largest duct where several are available because its larger dimensions facilitate placement of the unit. In addition, the higher air volume promotes a better evaporation rate and complete distribution. Nevertheless, if the smaller duct is the only alternative, it should

Humidity Control

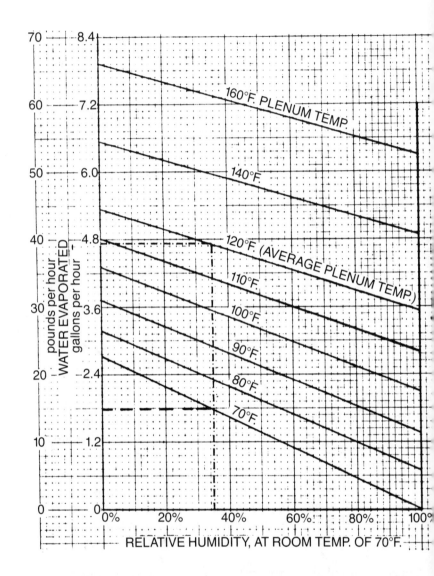

Figure 4-6. The output of an evaporative (wetted element) unit decreases with decreasing plenum temperature. The sample curves shown indicate that output drops from nearly 4.8 lb/hr at 120°F to only 1.8 lb/hr at 70°F.

Installation & Service

not be ruled out since the water vapor eventually works its way into the system—even if introduced in a remote section of a building.

Installation selection depends largely on the accessibility of: water supply, drain and electrical service. Follow and adhere to the manufacturer's instructions. Most manufacturers are cognizant of the relationship between proper installation and maximum performance. Consequently, they devote considerable time and effort in the preparation of installation manuals and service instructions. Not only do installation instructions suggest the best location for placement and service, they also include templates which service as an accurate guide for the placement of mounting hardware and required openings.

Another consideration to examine is locating the most effective position for a humidifier. Since it is a water handling unit, a humidifier relies on mechanical devices to maintain the proper water level. Although floats and water regulating devices are reliable, malfunctions can still occur and cause leaks and water spillage. The humidifier should be positioned so that water spillage or leaks cannot affect the furnace combustion area, oil/gas burners, or any electrical or mechanical equipment which can be damaged by water. In the event that installation sites are limited, pay careful attention and properly adjust the overflow connection to assure trouble-free drainage. In high rise or multi-story apartment buildings, for instance, accidental water spills can create thousands of dollars damage to ceilings, walls and furnishings. Subsequently, tubing in an addition drain pan may be an inexpensive form of insurance.

The important leveling procedure is also associated with the water problem. In bypass units where a manifold distributes water to evaporator pads, accurate leveling is critical to insure 100% pad saturation. In other designs, leveling insures proper water flow through the overflow mechanism.

Once an opening is made in a plenum or a duct, mounting the humidifier may require the use of a mounting frame, a support bracket, or merely fastening the unit to the plenum duct with sheet metal screws.

Humidity Control

Prior to mounting the unit, however, make certain that proper and adequate air flow will pass through the unit. Especially with duct mounted units, manufacturers often provide a baffle which is used to divert air to and through a humidifier. If a baffle is not provided by the manufacturer, improvise one from the sheet metal cut-out removed from the humidifier opening.

When positioning the humidifier, make certain the gaskets or sealing mechanism provided by the manufacturer is applied correctly. After positioning the unit, water supply and drain, complete the electrical connections.

Proximity of the water and electrical line is an important factor in positioning a humidifier. Generally ¼" copper or plastic tubing is adequate for a water supply line. Most manufacturers provide a self-piercing saddle (line tap access) valve which facilitates tapping into to an existing water line as illustrated in Figure 4-7. Once the valve is seated properly, run the tubing overhead to the humidifier supply solenoid or float valve.

On newer units, manufacturers recommend using ½" or ⅝" PVC tubing for the drain and overflow connections. A garden hose fitting or adapter can also be utilized on some units. In a non-critical drain and overflow application, the hose can be routed to the nearest floor drain. However, in a critical application, where water leakage or spillage can damage building construction or furnishings, the drain and overflow must be properly sized piping or tubing, permanently connected to a suitable drain system.

Another installation requirement which is becoming more prevalent is compliance with local and national plumbing and electrical codes. Plumbing codes are becoming very critical concerning float and water feed mechanisms used on humidifiers. The main problem is the possibility of back-syphoning contaminated water into the potable water supply. As a result, some major municipalities insist on humidifier acceptance by an established testing agency before the unit is permitted to be installed. However, an anti-syphoning device can be added to many humidifiers whose feed mechanisms which are not designed to prevent back-syphoning.

Installation & Service

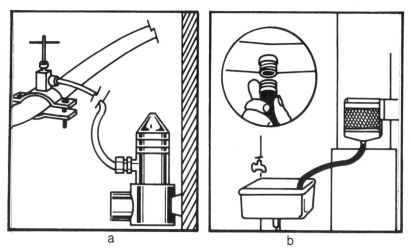

Figure 4-7. Self tapping valve installation: a) Connect self piercing saddle valve to the nearest unsoftened, cold water pipe. Follow instructions on valve packet and connect ¼" OD tubing to valve and turn on water for a few seconds to purge dirt or contaminates from the line. b) Connect other end of tubing to humidifier solenoid or float valve. Next, connect drain hose and run to the nearest suitable drain (floor, tub, sump pump or air conditioning condensate pan). Make certain drain line has a gradual slope.

Electrical & Control Systems

The electrical system on residential and small commercial humidifiers is relatively simple. Nevertheless, it must conform to local and national codes. Articles 110-2 of the National Electrical Code states: "The conductors and equipment required or permitted by this code shall be acceptable only when approved." The key word "approved" means labeled and listed by a nationally recognized testing organization such as the Underwriters Laboratories (UL) or the Canadian Standards Association (CSA).

Humidity Control

Before attempting any electrical work, make certain a power supply disconnect to the humidifier is provided. Also verify that the disconnect is "open" (i.e., the line is not energized) while wiring the humidifier power and control circuit. Remember, as little as 0.02 of an ampere can be fatal, and it can come from a low voltage (25V) as well as from a line voltage source.

The electrical system of a humidifier is a 120V power circuit which operates a motor or energizes a water solenoid valve or heating element. The control system determines the humidifier cycle rate either directly or through relays. The control system for a power humidifier is relatively simple. Actually, a humidifier can be installed with no humidity sensing device whatsoever (Figure 4-8). With this arrangement, the humidifier cycles whenever the furnace is on and the blower is operating. Although this system may be adequate under certain conditions, it is not equipped to adapt to the variable needs of an entire season. In extremely cold weather, for example, window condensation may create damaging water run-off. The only remedial action in this case is to cut-back or stop the humidifier.

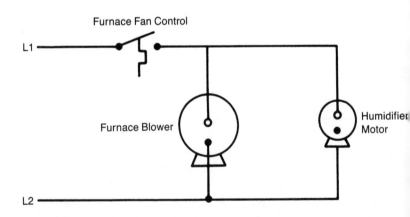

Figure 4-8. Schematic showing humidifier installation wired so the unit cycles whenever the furnace blower operates. (Reprint by permission—ASHRAE Handbook and Product Directory)

Installation & Service

The best method of controlling the humidifier is through a humidity sensor (humidistat). As humidity changes, the "live" element of a humidity sensor (human hair or nylon strand) lengthens and shortens to actuate a switch. The humidistat can be mounted on a wall or on the surface of the return air duct or plenum (Figure 4-9). The wall mounted unit provides better control because it is in the occupancy area. However, the return air humidistat provides adequate humidity control and utilizes a more simplified wiring job than that required for a wall mount.

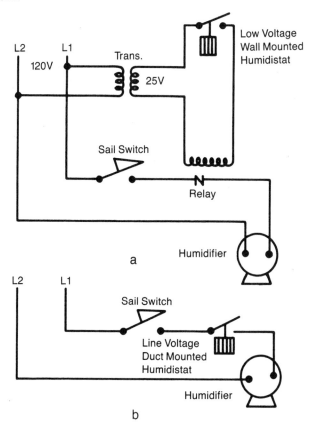

Figure 4-9. Schematics illustrating a) the use of a wall mounted humidistat to control humidifier operation, b) duct mounted humidistat which simplifies the required wiring.

Humidity Control

In a conventional control system, the power lead for the humidifier is taken from the load side of the fan control. The temperature actuated fan control only completes the electrical circuit to the blower when the furnace is capable of providing warm air at the registers. This interlocks the humidifier with the furnace blower so both operate together to assure that humidifier operation only takes place when a flow of warm air is available. The humidistat is then wired in series with the humidifier motor, relay, or in the case of a bypass unit, in series with the water solenoid valve.

Although the easiest way to interlock the humidifier to the furnace blower is through the furnace fan control, there are applications where this is impractical. Electric furnaces may have 240V furnace blowers and no thermal fan control. The blower is cycled with a time delay mechanism after the thermostat calls for heat. This presents an impossible problem of interlocking a 120V humidifier in the circuit because both motor leads must be hot in order to provide 240V. Although an electrical connection from the motor blower power lead to a neutral wire would provide 120V, it results in an uninterruptable circuit. If wired this way, the humidifier would operate constantly instead of cycling with the blower.

There are several solutions. One is to use a combination humidistat and sail switch in the return air duct. The sail mechanism does not allow the humidifier to operate unless there is sufficient air movement through the distribution system. When air movement is established, a contact closes to complete a circuit to the humidistat. The humidistat can then cycle the humidifier. If a wall-mounted humidistat is used, an individual sail switch can be used in the return air duct to move air flow through the humidifier. When this is established, the wall-mounted humidistat cycles the humidifier, the water solenoid valve or a relay. If a wall-mounted humidistat is mounted some distance from the humidifier, it may be practical to utilize low voltage wiring (Figure 4-9a). A transformer is used to step the 120V down to 25V. Inexpensive, unshielded light-duty wire can then be used to connect the humidistat to the relay. The lighter wire is also easier to run.

Installation & Service

Two-Speed Blower Operation

Due to different air requirements for heating and cooling, furnace manufacturers often utilize multiple speed blower motors—especially on high efficiency furnaces. The blower motor is equipped with a low speed winding for winter (heating) operation, and higher speed winding for summer (cooling). Figure 4-10 illustrates the burn-out hazard if a power humidifier is incorporated in the circuit without taking precautionary measures. There are two types of single phase, two-speed motors: multi-pole (split phase or capacitor start) or tapped winding, variable torque (permanent-split capacitor or shaded-pole). Regardless of type, a transformer action can take place between motor windings if the humidifier motor winding completes a circuit with the low speed blower motor winding, while the high speed motor winding is energized. This creates an auto-transformer action between the energized high speed winding and the completed circuit of the low speed winding and the humidifier.

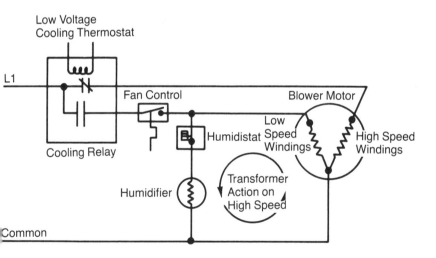

Figure 4-10. A potential burn-out hazard is present if a power humidifier is connected to a circuit without using proper precautions. (Reprint by permission — ASHRAE Handbook and Product Directory)

Humidity Control

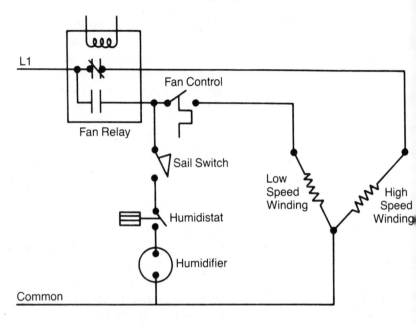

Figure 4-11. Schematic showing how a humidifier is connected to the supply side of a fan control to isolate the low speed windings.

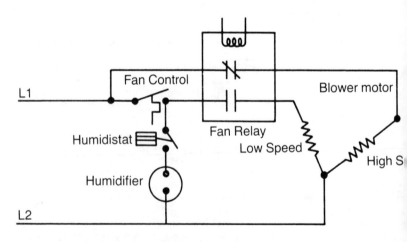

Figure 4-12. Alternate method of isolating the low speed motor winding by using a double-pole, single-throw fan switching rely.

What results is increased current draw and heating of the motor. In addition, the completed circuit in the low speed winding tries to operate the motor at low speed which works against the high speed operation. Again, this creates increased heating. Eventually a burn-out occurs.

In order to avoid a burn-out, the low speed motor winding must be electrically isolated from the humidifier or any device which provides a complete electrical path. This prevents transformer action when the low speed winding is open. One way to isolate it is to use a sail switch instead of the fan control to prove air flow. This allows the humidifier to be electrically isolated from the low speed motor winding through the open fan control.

In Figure 4-11 the humidifier is connected to the supply side of the fan control. The open fan control isolates the low speed windings and no transformer action takes place. The humidifier can only operate during the heating cycle when the sail switch senses air-flow and the humidistat calls for a cycle.

Figure 4-12 shows another method of isolating the low speed motor winding. The usual single-pole cooling fan relay is replaced by a double-pole, single throw fan switching relay. When cooling is required, the fan relay breaks the circuit to the low speed winding and makes the circuit to the high speed winding.

With an open circuit to the low speed winding, there is no way to induce voltage by transformer action. Consequently, the low speed winding is completely isolated and safe. The humidifier can be interlocked to the heating cycle of the furnace blower by connecting it to the load side of the temperature activated fan control. In this mode, it only operates during the heating cycle when the blower operates and the humidistat is "on."

Again, the one overriding precaution required with two speed blowers is to make certain the low speed winding is completely isolated during high speed blower or cooling operation. The humidifier should never be in the circuit so the low speed blower motor windings have a complete parallel path through the humidifier motor windings.

Humidity Control

In summary, the control circuit for a humidifier, operated in tandem with a warm air furnace, is relatively simple. It is generally a complete 120V circuit unless a wall-mounted humidistat is utilized at a considerable distance from the humidifier. In this instance, it is advisable to operate the humidistat on a low voltage (25V circuit) which does not require shielded or enclosed wiring. A relay transformer can provide the low voltage and required switching to energize the humidifier when the humidistat calls for a cycle.

The humidifier circuit is typically wired in parallel with the motor blower. Before either can operate, the thermostatic fan control must be actuated. This guarantees the air movement and heat energy required by the humidifier to develop its full rated capacity. If it is inconvenient to connect the humidifier to the load side of the fan control, it can also be wired to a sail switch or pressure differential control in the duct.

Auxiliary Circuits

In addition to operating the humidifier motor or solenoid, the control system may be required to perform other functions. One, for example, is the purge cycle used to flush impurities out of the humidifier sump or reservoir. This control circuit must have a timing mechanism that deactivates the humidifier and energizes a solenoid valve which is installed along the drain line or a pump located in the reservoir. After completing the purge cycle, the timer returns control of the humidifier to the humidistat.

Another auxiliary control function exists where a pump picks up water from a sump and delivers it to a header so the water can trickle back down through pads facilitating vaporization. Although these auxiliary functions are relatively simple, they are necessary to either eliminate residue and minerals from the humidifier or to prevent water waste by recirculating it for another pass through the vaporizing media.

Installation & Service

Start-Up & Check-out Procedure

After installing a humidifier and all plumbing, mechanical and electrical connections have been made, cycle the unit several times. Testing the unit verifies that it functions as designed and can prevent costly call-backs which can be attributed to a minor malfunction. The instructions outlined in the manufacturer's instruction booklet usually presents a satisfactory start-up procedure checklist. One common recommendation is to open the water valve and check for leaks. If a float valve is used, make sure to properly adjust the water level in the reservoir pan to prevent any overspill. If the level is too high or low, either compensate with the adjustment screw or bend the float arm as directed in the instructions to correct the situation.

Once the humidifier's electrical connections are complete, start the furnace. When the blower starts, cycle the humidifier with the humidistat to determine that all parts are functioning properly and that adequate air is flowing through the unit. A detailed check-out procedure, typical of most that are included in manufacturers instructions, follows:

1. Inspect mounting of the unit. Make certain it is level and secured according to manufacturer's instructions.
2. Verify that no restrictions exist in supply piping and drain.
3. Inspect wiring to determine if it complies with manufacturer's recommendations and local and national codes.
4. Set disconnect switch to "on" position.
5. Turn humidistat to a high position so it calls for humidity.
6. Open saddle valve or gate valve to supply water to the humidifier reservoir.
7. Inspect operation of the solenoid valve or float valve to ensure that water is fed to the distribution system or to the reservoir (in accordance with unit design).
8. Check all water connections for leakage, and allow reservoir to fill. Monitor the flow rate if a float valve exists.

Humidity Control

9. Check water level in reservoir and make any necessary adjustments. If a solenoid is involved, check for proper water distribution through the media. In the case of a spray atomizing humidifier, nozzle should be directed according to manufacturer's instructions.
10. Check fan control and verify whether humidifier stops when furnace blower motor stops.
11. Inspect motor and controls and determine if they are functioning properly. Monitor motor current draw with an ammeter to confirm it is within specifications.
12. Determine if the proper amount of air is passing through the unit. If not, provide proper baffling as recommended by the manufacturer to divert required air flow.
13. Check mounting and duct to make certain no air is leaking to the outside.
14. If a central cooling unit is installed in the furnace, make certain the humidifier does not operate during the cooling cycle.
15. Set humidistat to correct design setting.
16. After installation, clean-up the job site and present a tidy, professional looking job to the customer for inspection.

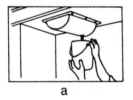 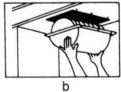 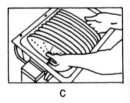

a   b   c

Figure 4-13. Cleaning procedure: a) disconnect power, shut off water supply, drain unit and disconnect water line, b) lower unit from duct or mounting, c) clean and/or replace wetted element, reservoir or its liner (as recommend by manufacturer) then ro install and assemble unit.

SERVICE PROCEDURES

Customer Education

The most important step in preventive maintenance is customer education. The most effective way to educate a customer is to simply explain the performance possibilities and limitations involved in humidification.

Typically, a humidifier is not a totally automatic, self-cleaning mechanism—even one equipped with bleed-off and flushing mechanisms. Moreover, water poses one of the most troublesome problems to mechanical equipment due to the large quantity required for proper humidification. The inherent mineral and chemical contaminants present in water promote chemical reactions, create wear and deposit build-ups (scale) on any metal surface if left untended over long periods of time.

Consequently, if the homeowner does not practice preventive maintenance by periodically inspecting and servicing a humidifier, a malfunction due to mineral build-up and sludge is certain to ensue. When the humidifier is dismantled after breaking down, it can look like a disaster area. Homeowners must realize that a humidifier requires service if it is to operate efficiently and vaporize hundreds of gallons of water. Like the vapor it produces, the benefits provided by a humidifier are invisible. However, the mess created by a neglected humidifier is painfully obvious. Be certain to emphasize this point to the homeowner. In addition, take a few minutes to point out the key maintenance points included in the manufacturer's instructions—especially motor oiling procedures. The extra effort will be well worthwhile.

Common Problems

The most common humidifier problems involve water impurities. Every water sample, other than rain, distilled and demineralized water, contains these impurities in varying concentrations. Distilled or demineralized water vaporize and

Humidity Control

leave or no residue, but they are not economically practical for use in home humidification.

As water is heated to form vapor, the mineral content of the water (especially lime—the prime constituent of scale) increases in concentration. When the mineral content of the water reaches the saturation point, the water can no longer hold the dissolved solids. Eventually the mineral solids precipitate out of suspension and solidify into a solid mass which cannot be easily be flushed out of a system. This, in turn, creates a difficult cleaning problem which can become insurmountable if not discovered in time. The speed at which this occurs depends largely on the hardness of the feed-water. The hardness of the water is usually measured in grains/gallon.

The grain hardness of water can be determined by consulting the local water department. The only solution to the water impurity problem is periodic draining and thorough cleaning of a humidifier.

When first placed in service, drain a humidifier after the first week of operation. After this initial purging, refer to the following schedule as guide for periodic draining:
1. 0 to 5 grains/gallon (very low hardness) drain once per heating season
2. 5 to 10 grains/gal (low hardness) drain twice per heating season.
3. 10 to 20 grains/gallon (average hardness) drain four times per heating season
4. 20 grains/gallon and above (very high hardness) drain once each month during the heating season.

Humidifier Cleaning

If mineral deposits have accumulated beyond ⅛" thickness, it is recommended to clean each component of the humidifier which comes in contact with water. This is apt to necessitate the removal of the unit from its mounting. After removal, if motor driven, set the unit some place where its operation can be tested. Start by filling the reservoir with water to normal capacity and add two pints of household vinegar or

Installation & Service

regular residential humidifier cleaner into the reservoir. Muriatic acid is an especially effective humidifier cleaner because it has a natural affinity for dissolving lime scale deposits. Muriatic acid is a weak solution of hydrochloric acid. However, since it is an acid, it must be handled with care.

> Note: Full strength muriatic acid is corrosive and can cause burns if it comes in contact with skin or clothing. If an accidental spill occurs, quickly and thoroughly wash the acid off. Always wear adequate eye protection, and if splashed in eyes, flush thoroughly with water and consult a physician.

If a humidifier is motor driven, it is a good preventive maintenance measure to oil the motor according to the manufacturer's instructions. In the case of a spray atomizing humidifier, be sure to clean the nozzle and water filter near the solenoid valve regularly.

Absence of Humidity

The "no humidification" or "low humidification" service call is perhaps the most common and the most annoying complaint. On such a call, a serviceman must systematically localize the problem. This is accomplished by following the planned check procedure suggested by the manufacturer. However, if the unit is completely inoperative, the first step is to determine if power is available to the unit. This entails checking the fuse or circuit breaker—before initializing other routine checks.

If power is available, then check the controls. Verify whether voltage is available from the fan control or sail switch to the humidistat. Make certain the humidistat is set high enough to make a circuit to the motor, solenoid or relay. Check all possible control malfunctions step-by-step and eliminate each until the problem is localized to one component. The motor is frequently the culprit and is easily replaced.

Humidity Control

If the unit is operational and the complaint is no humidity or low humidity, the first step is to determine the accuracy of the complaint. A homeowner may base his complaint on a personal perception or the readings indicated on a inexpensive, inaccurate hygrometer. Both can have considerable error. Utilize an accurate instrument type psychrometer and take readings in the most frequently occupied rooms in the house. If the low humidity reading can be documented, then take the following steps:
 1. Check the setting on the humidistat.
 2. Check the feed water system to determine if the humidifier has an adequate water supply.
 3. Inspect the home for unusual leakage losses such as undampered fireplaces or loose fitting doors and windows.
 4. Check the running time of the furnace blower and air temperature. The humidifier can only operate when the blower is running. If the blower has intermittent operation with long "off" cycles, the humidifier cannot operate. Also, low air temperatures diminish humidifier capacity. Set fan to constant "on."
 5. Determine the humidifier load requirement. If the humidifier is undersized, a larger unit or a second unit may be required.
 6. Verify whether sufficient air is properly directed through the humidifier. Install or adjust the baffle to direct air flow through the humidifier.
 7. Inspect the humidifier's wetting media (pads or screens) to determine if they are clogged and hampering satisfactory air/water contact.

Excess Humidity

This is by far the easiest type of complaint to solve. In most instances, merely setting back the humidistat or throttling down the unit can decrease the humidity in a house. Under capacity, on the other hand, is very difficult to correct if marginal sizing is involved.

Installation & Service

If an over-humidification condition must be corrected, refer to the following guidelines:
1. Determine whether a humidistat has been properly installed in a system.
2. Determine if the humidistat is set properly.
   - If humidification causes water run-off on the windows, throttle back the humidistat.
   - Test humidistat operation by running it up and down to determine if it controls the motor, solenoid or relay.
3. Check wiring and other controls. Determine if there is a short in the wiring or if a sticky relay has created a constant ON condition.
4. Check internal water circuit to determine if leaks are creating additional evaporating surfaces.

Air Noise

Another common complaint concerning humidifiers is noisy operation. This is not difficult to localize and is often caused by a worn, noisy motor or worn bearing surfaces which are easily pinpointed.

Air noise can also result from high velocity air directed through the humidifier media. Often merely readjusting the baffling solves this complaint.

SUMMARY

Service calls are expensive to both the contractor and the homeowner. This cost can be minimized by utilizing an organized approach. The best method—follow a maintenance checklist such as the one recommended by the ARI (refer to Figure 4-14). It is quite easy to customize a checklist for a specific humidifier by referring to the procedure recommended by the manufacturer.

A checklist is valuable in establishing an organized routine for service personnel. It can conserve time and eliminate wasted motion. In addition, using a checklist provides a documented service outline which the homeowner can review.

Humidity Control

You now have the essential fundamentals necessary to acquire a hands-on, working knowledge of residential humidification. The basic concepts presented herein, in addition to manufacturers' specific instructions, have equipped you with the proper "tools" to professionally design, install and service humidifiers for even the most demanding and critical residential customer.

Customer_____ Address_____

Equipment Location_____ Contract No._____

Date Last Inspected_____ Date This Inspection_____

Inspected By_____

System (Brief Description)

_____

Inspect, Check, Clean and Adjust, When Necessary, All items Listed

_____

EVAPORATING MEDIA OR ATOMIZER . . . .
    Clean or replace media, as required, or clean all
    parts, if atomizing type.
SUMP AND DRAIN LINES . . . .
    Clean sump and drain lines and make sure drains and
    drain lines are free of restrictions. Check for leaks and
    proper water level.
FANS AND MOTORS . . . .
    Fan, check for deterioration and balance. Fan set-
    screw, check for tightness. Lubricate motor bearings
    as required.
CONTROLS . . . .
    Water flow, check for proper flow rate. Humidistat,
    check for proper operation. All electrical terminals for
    tightness. Disconnect switch, check operation.

_____

Key: √—OK,   X—Need Additional Service,   XX—Repair or Replace

_____

Remarks:_____

_____

Figure 4-14. Typical humidifier preventative maintenance and service checklist.

SECTION 5

COMMERCIAL, INDUSTRIAL, INSTITUTIONAL HUMIDIFICATION

The primary purpose for residential humidification is the health, comfort and well-being of occupants. Essentially, humidity levels must be maintained to minimize body heat loss (due to evaporation), the drying of nasal passages and annoying static electrical discharges. Secondary reasons include the protection and preservation of hygroscopic furnishings and furniture, as well as the general building construction and, to a lesser extent, energy conservation.

Low residential humidity promotes widening of window and door cracks and, in turn, increases infiltration and exfiltration. Adequate humidity levels, however, minimizes this heat loss created by leakage. In addition adequate humidification permits a lower thermostat setting which also helps reduce heat loss. Although energy is required to vaporize water, a lower thermostat setting can often times provide a net energy saving which is sufficient enough to help offset the energy cost.

COMMERCIAL & INDUSTRIAL HUMIDIFICATION

Like residential humidification, commercial, industrial and institutional humidification are also concerned with the health, comfort and well-being of its occupants. However, concern for the well-being of occupants in the latter case may be due to a slightly different motivation than that involved in residential applications. For instance, if an improved working environment is provided by properly maintained air conditions, it can potentially enhance individual productivity in the work place. As a result, humidification can prove to be cost effective. Studies show that healthier working environments have also proven to reduce absenteeism which is a significant concern in most businesses. In addition, controlled humidification improves product quality, increases energy efficiency and eliminates static electrical charges which can be detrimental to some types of manufacturing operations.

Occupancy Comfort Levels

One of the most subjective, value judgements made in connection with environmental surroundings involves indoor comfort levels. Many individuals are influenced more by psychological than physiological factors. This means that even the color of walls can act as a stimulus to evoke an individual's reaction to the temperature of a room. This sort of reaction happens despite the fact that room temperature and humidity are at prescribed norms. Documentation exists that indicates "no heat" or "low heat" complaints in offices can been minimized or entirely eliminated merely by utilizing wall coverings (paint or wallpaper) with warmer colors.

Complaints of "no heat" or "low heat" usually originate from relatively few occupants in a building. Still, these complaints can create many problems for building maintenance personnel. In trying to appease complainants, who typically represent a minority of a building's occupants, maintenance personnel can easily over-compensate when adjusting the heating level. This can annoy the majority of the occupants who were comfortable at the original heating level. There is no easy remedy for this type of problem. One valid approach to comfort level complaints, however, is to adhere to researched and studied standards developed for indoor environments. Figure 5-1, developed by ASHRAE, is an acceptable winter/summer comfort chart for persons in typical summer and winter clothing who are engaged in primarily sedentary activities (typical of an office setting) and exposed to low velocity air movement. Attempt to educate occupants, who lodge "no heat" complaints, and make them aware of established comfort standards. Be forewarned, that trying to appease complainants instead of educating them tends to compound the problem rather than solve it. Remember the old adage: You can't please all of the people all of the time.

Building engineers and maintenance personnel should accept the responsibility of fine tuning a humidification control system so that it conforms as closely as possible to the parameters outlined in the comfort chart. Once this task has been accomplished effectively, it can enhance the productivity and performance of nearly every building occupant.

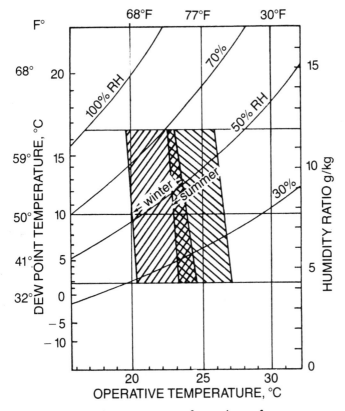

Figure 5-1. Winter/summer comfort chart for occupants in typical winter/summer clothing who are engaged in light work and are exposed to low velocity air movement. (Reprinted by permission—AHSRAE Fundamentals)

Humidity Control

TEXTILE

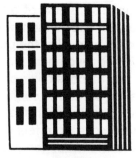
COMMERCIAL OFFICES

PRINTING

MEDICAL

WOODWORKING AND FURNITURE

COMPUTERS

Figure 5-2. Fields utilizing hygroscopic material which requires humidity control. (Reprinted by permission — Herrmidifier Company, Inc.)

Health & Humidification

Proper humidity levels minimize air borne bacteria and dust. In fact, properly moistened nasal and throat passages serve as a barrier to air borne bacteria. Documented medical studies indicate that maintaining a relative humidity between 35% and 50% in occupancy areas substantially reduces susceptibility to respiratory infections and colds. Consequently, in a commercial or industrial setting, proper humidity can potentially improve a worker's performance and reduce absenteeism due to illness.

Hygroscopic Material

The term hygroscopic refers to the characteristic of a material to absorb and retain moisture from the atmosphere. Hygroscopic material, then, is material that constantly absorbs moisture or rejects it depending upon ambient environmental conditions. During the process of water retention and loss, the chemical make-up and electrical nature of hygroscopic material constantly changes. Although these changes are slight and subtle, they must be reckoned with and controlled in product processing and preservation. The hygroscopic nature of a material takes on added importance because every material, except metal, undergoes significant changes with moisture variation—especially changes in physical dimensions.

Figure 5-2 illustrates many of the fields involved with hygroscopic material which requires controlled humidification and dehumidification. As a matter of historical fact, the humidity variation problem involved with the color printing process helped launch the air conditioning industry. In 1902, Willis Carrier confronted this problem at the Sackett-Wilhelms Lithographing and Publishing Company of Brooklyn, New York. His organized study of humidity and air interaction provided the basis for a controlled environment which solved the printing problem. More importantly, Carrier's work in this case led to the development of the psychrometric chart which yielded quick and accurate reference for many air handling problems.

Humidity Control

Carrier determined that controlled humidity during the printing process eliminated paper curl caused by expansion and contraction of uncoated paper. In addition, he found that color register, ink drying, static cling and feeding/binding production problems could be minimized or eliminated by proper humidity control. The humidity parameters he defined can vary from 40% to 60% RH depending upon the process being considered. Dry bulb temperatures must be maintained between 75° to 80°F.

Like paper, the amount of water wood can retain is substantial. A cubic foot of wood weighing 30 lbs, for example, can hold more than 3 pints of water at 60% RH. When the RH is lowered to 10%, which is common in unhumidified winter environments, the moisture content can drop to less than 1 pint. This dramatic decrease in moisture content results in dimension changes such as shrinking, warping, checking and splitting. In wood working, furniture manufacturing and even in bowling lanes the liveliness and uniformity of wood can only be maintained if its internal moisture content is kept relatively constant.

Leather is similar to wood in that it requires proper humidity to prevent deterioration and to retain its liveliness or pliability. Many volatile solvents are employed in the leather tanning process. As a result, 50% RH or higher is recommended in order to prevent static discharges and reduce the hazard of explosions.

Computer rooms, data processing and telecommunication operations represent specialized applications where controlled humidification and dehumidification is mandatory. Theoretically, solid state electronic components used in computers and for telecommunications can function and last forever since there are no moving parts to wear out. Nevertheless, these same components can be destroyed in an instant if subjected to excessive temperature or power surges. Static electricity discharges can also destroy components if measures are not take to prevent the charges. Consequently, a minimum of 40% RH is recommended in environments where use or manufacturing of solid-state components are encountered, and a range of 45% to 50% is preferable. This range also maintains quality and dimensional stability for circuit boards and magnetic storage media such as tape or disks used in data processing. (Tape brittleness is minimized with adequate humidity.)

From the standpoint of preservation and the elimination of weight loss, humidification can prove cost effective in the processing and storage of fruits and vegetables. Most vegetables, with the exception of onions, can be subjected to humidity levels approaching 100% RH without exhibiting decay. For example, in tobacco processing it is necessary to maintain high humidity levels from 60% to 65% RH in order to produce a quality product. Meat and poultry processing, on the other hand, presents another area where controlled humidity prevents shrinkage and product discoloration.

Static Electricity

One of the problems that result from inadequate humidity is static electricity. Its presence, however, can be more than just a annoyance or nuisance. As previously indicated, it is destructive in solid-state and potentially flammable environments. In addition, static charges can create degradation problems in photography and x-ray development rooms. Because of friction, static charges build-up and cannot dissipate if humidity is insufficient. Humidity provides a grounding path for static charges. The magnetic attraction created by static build-up can seriously interfere with automated feed and sorting mechanisms that handle paper, cellophane and plastics. The solution, maintain a RH of at least 50% (ASHRAE recommends a minimum RH of 55%).

HUMIDIFICATION LOAD CALCULATIONS

Before proper humidifying equipment can be selected, the humidification load must be calculated. Not only must the load be calculated, but it must be done with accuracy and attention to every variable whether pertinent to the calculation or not. If load calculations are inaccurate or are done haphazardly, under- or over-capacity in equipment sizing and selection is likely to result. Either condition poses a problem. The problem of under-capacity is obvious when the humidification requirements of an application cannot be met at critical load periods. Over-capacity, on the other hand, creates a special set of problems related to poor and erratic control performance. This situation, in turn, results in inefficient energy use and complaints by occupants concerning comfort conditions.

Design Factors

Heating and humidification are directly related functions. Outside air constantly moves into a structure. In older buildings, this occurs due to inherent infiltration and exfiltration and is defined as the number of air changes/hour. However, no precise engineering formula can determine the exact number of air changes/hour. Rather, an estimation of the number of times the internal air volume is replaced by infiltration/exfiltration is made based on the construction of the building envelope. Most modern structures have central fan and air handling systems designed to regulate and control precise amounts of entering outside air. And these amounts are usually determined by local ventilation codes. Regardless of the method employed, however, outside air that exists at a temperature lower than that specified for indoor design conditions usually requires the addition of moisture. The amount of moisture required is calculated by determining the difference between moisture content of untreated outside air (M_1) and moisture content of indoor treated air at design conditions (M_2). The best source for locating this information is the psychrometric chart. The extreme right side of the ASHRAE psychrometric chart in Figure 5-3 shows the moisture content along the vertical column labeled humidity ratio. The humidity ratio represents moisture content in lbs/lbs of dry air. Other psychrometric charts may show moisture content as grains moisture/lb of dry air. And many charts show use both designations to facilitate conversion between the two.

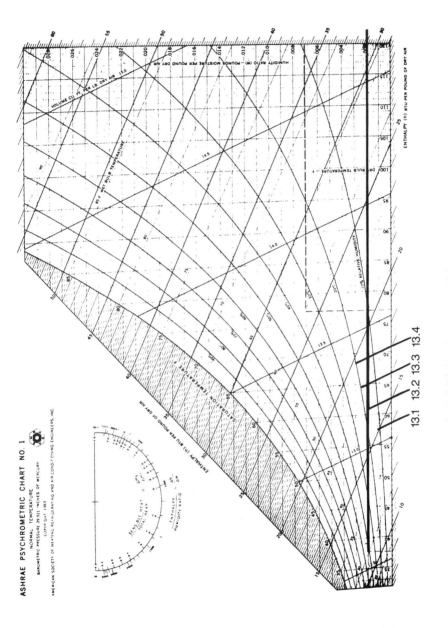

Figure 5-3. Psychrometric chart showing moisture content difference between untreated outside air and that of indoor air at design conditions.

Humidity Control

After determining moisture content between air at indoor design temperature and entering untreated outside air, then a calculation must be made to determine the pounds of untreated outside air entering the structure in a one hour period (D). This entails selecting the specific volume of outside air and calculating its reciprocal $1 \div V$. This establishes the density (lbs air/ft³) of the entering air. When the density is multiplied by the volume of entering air in one hour times the moisture content difference between entering air and indoor design air, then the moisture requirement H (lbs/hr) is established. This formula becomes the basis for equipment selection.

The following is the formula used for an application where outside air enters through infiltration and exfiltration:

$$H = D \times BV \times (M_2 - M_1) \times R - S + L$$

Where:
H = required lbs moisture/hr
D = lbs moisture/ft³ dry air ($D = 1 \div V$)
BV = building volume (area) to be humidified
M_2 = lbs moisture/lb dry air (indoor design specs)
M_1 = lbs moisture/lb dry air (outdoor design conditions)
R = air changes/hr
S = additional moisture — internal sources (occupants contribute 0.2 to 0.6 lbs/hr depending on activity, high moisture products in building)
L = hygroscopic materials that absorb moisture and add to load.

> Note: Values S and L are usually not considered unless they add to the load significantly.

The following is an example using the load formula:

$$H = 0.079 \times 100{,}000 \times (0.007 - 0.002) \times 2 = 77.9 \text{ lbs moisture/hr}$$

Where:
BV = 100,000 ft³
R = 2 air changes/hr
M_2 = 0.007 lbs/lb dry air (Figure 5-3) = @78° DB, 35% RH
M_1 = 0.002 lbs/lb dry air (Figure 5-3) = @39° DB, 40% RH
D = 1 ÷ 12.6

Note: specific volume of outdoor at 39°F, 40% RH = 12.6

Therefore the moisture necessary to maintain design indoor conditions is 77.9 lbs moisture/hr = 9.4 gph = 0.16 gpm.

Central Fan Humidification Calculations

The humidification calculation based on air changes per hour is an expedient method for use with structures unequipped for the introduction of specific amounts of outside air. This applies to most residential applications and many of the older commercial/industrial buildings as well as institutions. Modern structures, on the other hand, are required by code to provide a specific amount of outside air at all times. A central fan system introduces a calculated amount of outside air which is expressed as cubic feet/minute (cfm).

When using the psychrometric chart in load calculations, the formula is similar to the preceding example of the air change method except that BV must be replaced by ft³/hr (cfh) and the value R is eliminated. Consequently, the formula is:

$$H = D \times cfh \times (M_2 - M_1) - S + L$$

Where: cfh = cfm × 60

Note: values S and L represent internal moisture gains and losses.

An example of a central fan humidification load calculation for a system requiring 3,500 cfm of outside make-up air and with the same outdoor and indoor design conditions as the preceding air change application is:

$$H = 0.079 \times 210{,}000 \times (0.007 - 0.002)$$
$$= 82.95 \text{ lbs moisture/hr (10 gph)}$$

Where:
outside design conditions = 39°F, 40% RH
indoor design conditions = 78°F, 35% RH
cfh = 3,500 cfm × 60 = 210,000
D = density = 0.079
S, L = insignificant internal gains and losses

Charted Load Calculations

The preferred method of all hvac/r calculations is utilization of the psychrometric chart. It must, however, include the correct temperature range and altitude of the application in question. Still, some humidifier manufacturers employ quick reference charts which are based on values derived from a psychrometric chart. Table 5-1 is an example of the most commonly used chart. It differs from the psychrometric chart in this respect: it is based on 1,000 ft^3 of air (presumed to be air at standard sea level barometric conditions at 70°F) with a density of 0.075 lbs/ft^3. As a result, the specific volume or ft^3/lb of standard air is 13.33 – although many engineers use 13.5 as the standard. The psychrometric chart, on the other hand, is based on lbs moisture/1 ft^3 of dry air which demonstrate precise specific volume and density values. Therefore moisture calculations for identical applications may vary slightly, but not significantly, when using either chart. This statement can be verified by using the chart in Table 5-1 to calculate the previous example. The formula used for either chart is virtually the same. The only difference is that the chart in Table 5-1 lists the M_2 and M_1 values in lbs/1,000 ft^3 (presuming standard air density). However, the cfh value must be divided by 1,000.

Commercial, Industrial Humidification

Table 5-1. Total amount of vapor which 1,000 ft³ can hold at various temperatures and relative humidities.

Moisture Content* % Relative Humidity

Temp. °F	20	25	30	35	40	45	50	55	60	70	80	90	100
−10	.008	.010	.012	.014	.016	.018	.020	.022	.024	.028	.032	.036	.040
0	.012	.015	.018	.021	.024	.027	.029	.032	.035	.041	.047	.052	.059
10	.020	.025	.029	.034	.039	.044	.049	.054	.059	.069	.078	.088	.098
20	.031	.039	.047	.055	.063	.071	.079	.087	.095	.110	.126	.142	.161
30	.051	.064	.077	.089	.103	.116	.128	.141	.154	.180	.203	.232	.258
40	.077	.097	.117	.135	.155	.174	.193	.213	.233	.272	.310	.358	.390
50	.123	.141	.171	.198	.227	.259	.284	.312	.341	.399	.456	.521	.574
60	.162	.205	.246	.288	.329	.370	.412	.453	.495	.579	.663	.747	.831
62	.176	.220	.264	.308	.353	.398	.442	.486	.532	.622	.711	.801	.893
64	.189	.236	.284	.332	.379	.427	.475	.524	.572	.668	.764	.863	.960
66	.203	.254	.306	.356	.407	.458	.510	.561	.613	.717	.821	.926	1.030
68	.217	.272	.326	.381	.436	.491	.547	.602	.657	.768	.880	.991	1.106
70	.232	.291	.349	.408	.467	.527	.585	.645	.704	.824	.944	1.063	1.186
72	.249	.312	.375	.438	.501	.564	.628	.691	.755	.884	1.013	1.139	1.273
74	.267	.334	.401	.469	.536	.605	.671	.740	.809	.947	1.084	1.224	1.364
76	.285	.357	.429	.502	.574	.647	.720	.792	.866	1.012	1.160	1.310	1.461
78	.305	.382	.458	.535	.614	.692	.769	.848	.925	1.082	1.241	1.400	1.564
80	.326	.408	.490	.574	.656	.740	.823	.905	.991	1.160	1.330	1.500	1.676
82	.348	.436	.524	.612	.701	.790	.881	.970	1.060	1.240	1.422	1.609	1.792
84	.371	.464	.558	.654	.749	.844	.939	1.033	1.129	1.322	1.520	1.710	1.916
86	.389	.496	.597	.698	.799	.901	1.003	1.105	1.212	1.418	1.627	1.833	2.048
88	.422	.523	.635	.743	.851	.960	1.068	1.178	1.288	1.510	1.731	1.958	2.189
90	.450	.563	.678	.793	.908	1.024	1.141	1.258	1.375	1.613	1.853	2.095	2.338

*Pounds per 1000 cubic feet of air (based on standard conditions)

A calculation for 3,500 cfm of make-up air at 39°F and 40% RH, when humidified to design conditions of 78°F and 35% RH, is as follows:

$$H = (210,000 \div 1,000) \times (0.535 - 0.150) = 80.85 \text{ lbs}$$

Where:
$M_2 = 0.535$ lbs/1,000 ft³ (from Table 5-1)
$M_1 = 0.150$ lbs/1,000 ft³ (from Table 5-1)
　　[M_1 interpolated from values of 40°F, 40% RH and 30°F, 40% RH]

Humidity Control

Using the psychrometric chart, the humidification load calculation yielded a figure of 82.95 lbs/hr (10 gph), while the lbs/1,000 ft³ chart yielded 80.85 lbs moisture/hr (9.7 gph). The difference between the two results is equivalent to less than 3% and is of no great consequence unless the precision of the humidification calculation is a high priority. In this instance, the use of the psychrometric chart method is highly recommended.

Industry manufacturers sometimes use another formula to determine the lbs moisture to maintain indoor design conditions (value H) by using Table 5-1.

$$H = \frac{cfm(M_2 - M_1) \times C}{16.7}$$

This method is expedient as it incorporates the central fan system's established cfm or ft³/minute of capacity. The value C delineates (in percentage) how much of the system cfm is make-up or outside air. In effect, multiplying by the constant 16.7 makes the cfm to cfh conversion. It also effects a division by 1,000 in the following manner:

$$16.7 = 1,000 \text{ ft}^3 \div 60 \text{ min/hr}$$

Applying this conversion to values from the previous example yields the following:

$$H = \frac{35,000 \text{ cfm } (0.535 - 0.150) \times 10\%}{16.7} = 80.69 \text{ lbs/hr}$$

Nomograph Load Calculation

The quickest method of determining humidification load is done by using the nomograph method (Figure 5-4). It merely involves plotting V (volume of air entering the structure) on the left hand column and M (required moisture content in lbs/lb of dry air from the psychrometric chart) on the right hand column.

A line connecting these two points intersects the middle H column, and this gives the humidification load in both gph and lbs/hr.

Commercial, Industrial Humidification

The dotted line examples show solutions for 100% outside air conditions and for design outside air conditions where only a percentage of the total cfm is outside air. To determine the load with 100% outside air subtract the psychrometric chart moisture content for outside air from the value for indoor design conditions. Plot this on the M column of the nomograph. Next, determine the volume of air and plot this point on the V column. If the volume of air is greater than 10,000 ft^3, for example 50,000 cfm, use a value which is listed on the chart—5,000 cfm for instance (an equivalent to a multiple of 10). Then plot a line from a V of 5,000 to the calculated M value. The answer in the H column can then be multiplied by a factor of 10 to yield the solution for the 50,000 cfm application.

In design outdoor air conditions where only a percentage of the total air volume is outside air, the calculated M value can be multiplied by the percentage of outside air and then plotted on the M column of the nomograph. Then by plotting the total volume on the V column are drawing a line between M and V, the value for H is located at the intersection.

The solid line on the nomograph in Figure 5-4 is the previous example of a central fan system delivering 5,000 cfm with 10% or 3,500 cfm of outside air. The 3,500 cfm value represents the total load and can be plotted directly in the V column. With outside design conditions of 39°F, 40% RH and indoor design of 78°F, 35% RH, $M_2 - M_1$ is again 0.005 lbs/lb of dry air and is plotted in the M column. As indicated in Figure 5-4, by connecting V and M the H intersect occurs at approximately 82.9 lbs/hr (10 gph) which is identical to the solution calculated in the example.

Humidification load calculations must be accurate enough to select equipment to meet maximum requirements at the outdoor design conditions. With sufficient capacity assured, modern control technology can modulate and maintain even the most demanding humidification needs with utmost precision and reliability by utilizing direct digital control microcomputer capabilities.

Humidity Control

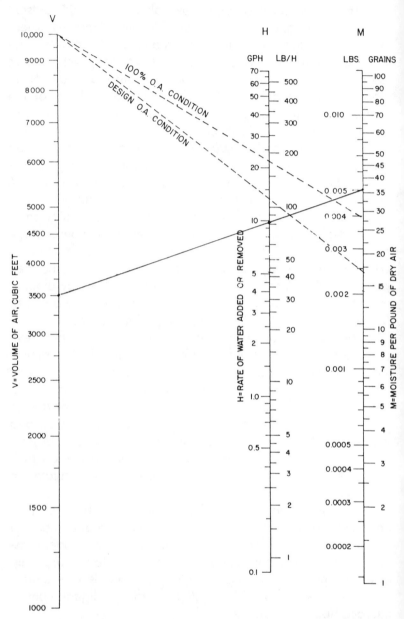

Figure 5-4. Nomograph used to determine humidification load. (Reprint by permission—Johnson Controls, Inc.)

COMMERCIAL & INDUSTRIAL HUMIDIFIERS

There are a number of effective humidification systems available for commercial and industrial applications. These primarily include:
- heated pan type
- atomizing type
- steam type
- infrared type

HEATED PAN HUMIDIFIER

The heated pan humidifier, Figure 5-5, is the simplest of all humidifiers. Basically it consists of a shallow pan which has a large surface area for evaporation. This type of unit is designed to be located in the air stream or mounted on the lower surface of ductwork. The heat energy source for vaporization is either an electric heating element or a steam/hot water coil immersed in the pan. The correct water level is maintained by a float valve, and the system is protected from electrical element burnout by a low water level cut-off.

It is important that moisture carry-over into the air stream be avoided. This condition occurs due to violent nucleate boiling and/or splashing. The most desirable condition occurs when nucleate boiling bubbles form at the water heating surface but dissipate in the upper water strata before reaching the air/water surface. The dissipation of the superheated bubbles superheats the liquid water which then rises to the surface forming a liquid-vapor interface for maximum evaporation. This process tends to eliminate splashing which minimizes the addition of sensible heat to the air stream. Splash prevention is accomplished by increasing the pan depth and water level, increasing the coil or element length, reducing steam pressure or energy input, or by adding baffle splash eliminators. Actually baffle splash eliminators must be used when steam pressure exceeds 15 psig.

The key advantage in the use of a pan humidifier is its low cost. However, there is a trade-off in that frequent maintenance is required to keep the pans free of lime and scale

Humidity Control

deposits that build-up and flake off due to thermal expansion and contraction caused by the heating element. The use of treated or demineralized water can minimize the problem of mineral deposits however.

Commercial demineralizers are very effective in eliminating minerals and dissolved solids which contribute to water hardness. After prolonged operation in the absence of a demineralizer, solid residue build-up in the pan reduces evaporation capacity. The concentration of dissolved solids can be diluted and removed from a unit by periodically purging the humidifier with a timed flush cycle which employs a liquid valve solenoid actuated by a time clock. In most cases, if maintenance personnel are readily available, purging can also be done manually with a hand valve. If residue tends to

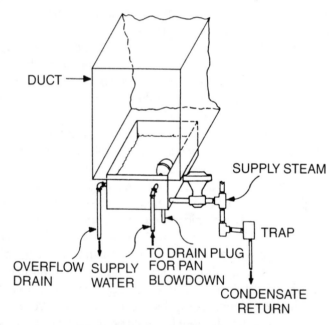

Figure 5-5. Typical pan humidifier installation diagram. Automatic make-up water control is provided by a built-in float valve located in the pan. Water in the reservoir is heated by an electric element immersed in the pan. (Used by permission—Johnson Controls)

stick to the pan lining over a prolonged period of operation, it can be removed by draining the pan, adding a 10% acid solution if inhibited sulfamic acid (HSO_3NH_2) and boiling until the surfaces are free of deposits.

Proper inspection and maintenance allows the pan type humidifier to be effective where the humidification load is stable and not overly critical. The pan type humidifier has its greatest limitation where humidification loads may change rapidly. The response time of the pan humidifier is typically too slow to meet the demands of precision controlled performance.

ATOMIZING HUMIDIFIERS

Water spray or atomizing humidifiers, Figure 5-6, create a fine mist which is introduced into the air stream of a central air handling system. The fine mist can be the result of a high speed disc, which slings water through a fine comb, or it can rely on water pressure produced by a specially designed nozzle. Another, the dual pneumatic type, uses a combination of air and water pressure to atomize water. The dual pneumatic type produces droplets so small that they vaporize rapidly in the air stream.

The vaporization in the atomizing process is fundamentally adiabatic in that no heat energy is gained or lost. Sensible heat is converted to latent heat. This results in a decrease in dry bulb air temperature with a subsequent increase in moisture content and relative humidity. The new conditions can readily be determined by plotting the known variables on a psychrometric chart. This shows that wet bulb temperature and total heat remains constant in the atomizing mode. However, the noticeable decrease in sensible heat and dry bulb temperature may require the addition of sensible heat to maintain the human comfort range as prescribed in Figure 5-1. This is especially true if unheated water is supplied to a humidifier.

The advantages of an atomizing humidifier: no external heat energy source is required and instantaneous humidity is supplied on demand. In addition, most atomizers can be modulated which lends them excellent controllability characteristics.

Humidity Control

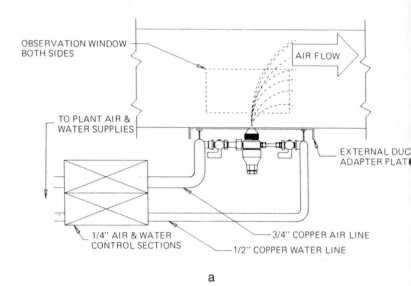

a

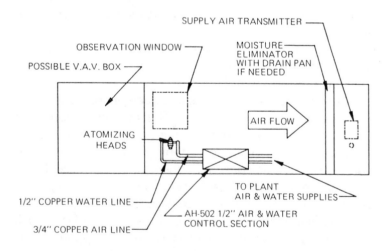

b

Figure 5-6. Two atomizing humidifier designs: a) dual pneumatic type, b) nozzle type. (Used by permission — Herrmidifier Company, Inc.)

The major disadvantages of the atomizing humidifier—the discharge and "dusting" of water soluble impurities and minerals in the conditioned air stream and the possible occurrence of moisture "fall out" in the duct work. These shortcomings can be minimized, although it involves supplemental equipment which adds some expense. Dusting is usually not a major problem—provided clean, relatively mineral free water is available. Unfortunately, there are very few areas of the country that can boast of this luxury. The solution is to use water demineralizers, air filter eliminators or both. Moisture "fall-out" is of special concern. In addition to ductwork rusting and deterioration, moisture fall-out results in algae build-up as well as odor and sanitation problems. The solution, proper design and application procedures. As extra insurance, the addition of a moisture eliminator and drain pan can be used as illustrated in Figure 5-6b.

STEAM HUMIDIFIER

There are many advantages in using steam humidification. First and foremost, steam is actually water vapor under pressure which can be immediately absorbed by air. The absorption process involves little or no lag, and this permits steam to have excellent controllability. Moreover, steam humidification is basically an isothermal process which simply means the temperature of the air remains constant—no significant increase in dry bulb temperature occurs. In addition, little mineral dust and/or impurities are introduced to the air. In a properly designed and engineered system, no water droplets are created, and subsequently, no corrosion, sanitation or odor problems result.

Another advantage in employing steam for humidification is that hospitals and many commercial/industrial buildings already have a source of steam available from central steam heating boilers. In this case, sufficient steam can be siphoned off for use in the humidifier. The trend towards the use of high temperature water, heat pumps and other heating methods has minimized and reduced the advantage and availability of steam boilers. Moreover, a steam boiler loses

Humidity Control

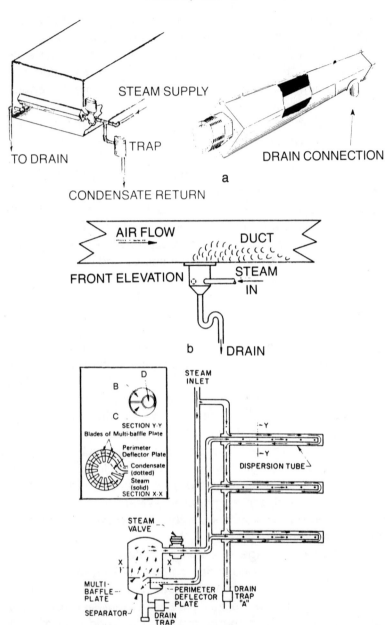

Figure 5-7. Steam humidifiers: a) enclosed grid type, b) cup or pot type, c) jacketed dry steam type.

Commercial, Industrial Humidification

its advantage if it is located a considerable distance from the air handler. Nevertheless, auxiliary steam generators and self-contained steam humidifiers are available for many steam applications. Alternate methods of steam generation, however, have the disadvantage of increased costs over other methods. In addition, steam pressure vessels are subject to strict codes and ordinances which govern their installation and use. Another disadvantage of steam use for humidification is the odor problem that results from boiler cleaning compounds.

- Enclosed Grid Humidifier—There are four basic types of steam humidifiers. The first is the enclosed grid humidifier, Figure 5-7a. Steam from a supply source enters a valve which modulates steam flow or operates as a two position valve controlled by a humidistat. Steam from an external supply source enters the valve, and condensate is drained to the trap and out to a drain. Steam enters the tube and is ejected through holes in the bottom of the tube. Capacity of the system varies with the perforations per linear foot of tube. This type of humidifier should be used with low pressure steam only in order to prevent condensate migration to the ductwork. The drain, located on the side opposite the control valve, eliminates moisture condensate as steam hits the walls of the enclosure. This drain must incorporate a drip leg and trap. Finally, the dry steam is discharged through a slot on top of the enclosure and to the passing air stream.
- Cup or Pot Humidifier—The cup type steam humidifier (Figure 5-7b) is a relatively simple device which attaches to the underside of the air handling system ductwork. The steam supply line or lines are tangentially attached to the inner periphery of the cup. Steam is ejected upward in the air-stream while condensate flows through a trap and out the drain.
- Jacketed Dry-Steam Humidifier—The jacketed dry-steam type humidifier is one of the most effective and the most popular of the steam humidifiers, Figure 5-7c. It utilizes a single or multiple dispersion tube and discharges steam directly into the air stream. When the supply service valve is opened it allows steam to preheat the dispersion tubes and

Humidity Control

to enter the water/steam separator. Steam and entrained condensate enter the separator. The condensate strikes perimeter deflector plates where it is forced to flow in a circular pattern. Gradually the condensate slows down until it exits through drain trap. The steam then rises through a multi-baffle plate. When the control steam valve opens, steam is allowed to enter the inner chamber or tube (D) and allowed to discharge into the air stream. Any condensate formed in this process is re-evaporated by heat energy from steam in the jacketed portion of the dispersion tube. This type of humidifier has the best design for eliminating condensate from entering the air stream. The trade-off, however, is that steam in the jacketed manifold condenses if cool air is supplied by the air handler. This must be compensated for by additional steam with no increase in capacity. Moreover, the dispersion tubes act as heating coils in the off-cycle, even though there may be no demand for heat.

- Area Humidifier—Area or self contained humidifiers that discharge humidity directly to the building area atmosphere are the answer when no air handling system exists. Even when air handling systems are available there are applications where supply air temperature is too low to absorb enough humidity to meet design requirements. In either event, steam humidifiers similar to the jacketed dry-steam humidifier can be located directly in the area to be humidified. Dispersion of steam to atmosphere is typically achieved by jet nozzle assisted by a circulating fan. Both the steam valve and the fan are cycled by the humidistat. If appearance is essential, manufacturers have area unit humidifiers which are mounted in attractive enclosures which can be floor, wall or ceiling units.
- Self-Contained Electrode Humidifier—The final type of steam humidifier is the self contained electrode type. This unit is the solution where no central or auxiliary source of steam is available. An added advantage is ease of installation due to prewiring and prepackaging. This feature allows a unit to be located in the most desirable proximity in relation to the air handler.

Commercial, Industrial Humidification

This type of humidifier generates steam via electrodes immersed in the water cylinder. Solid state control technology is utilized to detect mineral concentration and initiate dump or flush cycles. Electrode contamination is also detected and compensated for by increasing the water level in a unit. The disadvantage of this type, the electrodes must be replaced eventually, especially in areas where the mineral content of the water is high.

INFRARED HUMIDIFICATION

Infrared humidification is relatively new and very limited in its applications. Universally infrared is the most basic and generic of all humidification methods. More than 50% of the sun's energy is emitted as infrared radiation. As such, infrared has the speed of light, obeys the inverse square laws and, like light and radio waves, is electromagnetic. Therefore infrared has the ability to concentrate energy on distant targets with little intervening loss. A case in point is the sun, its infrared radiation spins the million miles to earth in a mere matter of seconds. In so doing it provides energy for the earth's most important humidification need—the hydrological cycle. Despite its transmission advantages, use of infrared for both heating and humidification has lagged well behind that of more conventional methods such as conduction and convection.

The one great advantage for infrared humidification is that water has a great ability to absorb rather than reflect infrared radiation. At thermal equilibrium, emission equals absorption and water gets no hotter. This characteristic allows the generation of low temperature, uncontaminated water vapor without promoting boiling action. Moreover, the infrared energy source and the vaporizing water are completely separate. Since no physical contact is necessary, the energy source is not subjected to contaminants. This significantly reduces maintenance costs involved with infrared energy equipment.

Figure 5-8 illustrates a very basic infrared lamp configuration. In practice, the infrared energy source can be provided by metal foil heaters or infrared power tubes in the form of high intensity quartz type lamps. Metal foil construction is

Humidity Control

based on the use of a special refractory insulating material and specially developed foils. When current is passed through the broad metal foil ribbons, they are heated above 1,500°F and the entire area glows uniformly. The radiant intensity produced approaches the maximum theoretical value for a source at that temperature. The heat loss from the back side is effectively blocked by the thermal insulating material on which the elements are mounted.

In the lamp application, the heat from the lamp filaments, operating at 4,000°F is focused and targeted by reflectors onto a water source provided by a pan. Water in the pan is maintained at a constant level with a float valve. The invisible rays of infrared energy flash pure water vapor from the surface of the water reservoir, and the vapor is carried off into the air conditioned space by the air handling system. The movement of air aids in the process by rippling the water surface. This rippling action accelerates humidification by diminishing water tension and preventing insulating films of contamination from forming on the surface.

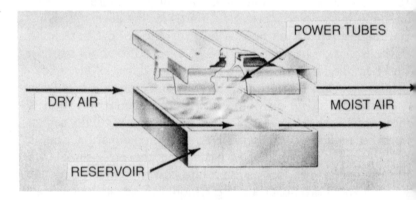

Figure 5-8. Basic infrared humidifier operation: infrared tubes flash water in reservoir into pure water vapor. Incoming air is directed in between the tube and water reservoir where it picks up vapor flashed from the surface of the water. Note, no carry-over of contaminates such as lime occurs to dust the space being humidified.

The infrared system suffers from the same limitation as all other humidifiers—the build-up of mineral and contaminant residue inherent in vaporizing water. Just as in the pan humidifiers, the infrared pan should be flushed periodically and automatically with a solenoid "dump" valve and time clock. However, the dump cycle does not eliminate periodic manual clean up and maintenance. Still, the automatic dumping does minimize the need for manual maintenance and extends the interval between required clean-ups. Despite this limitation, infrared humidification certainly does exhibit controllability advantages. When the humidification sensor initiates a cycle, the system is instantly activated.

When the sensor set point is reached, it provides immediate system shut-off. The fact that an infrared system has no inherent leads or lags in operation, and its energy system can be proportioned with solid state devices, lends itself to providing precision performance.

HUMIDIFICATION SYSTEM SELECTION

After total humidification loads have been calculated, the next step is system selection and design. The factors involved in this process follow:
1. Client's input or design preference.
2. Designer's judgement and preference.
3. Availability and characteristics of water supply, steam supply, electrical service and drains.
4. Space available in central fan system, system ductwork or packaged air conditioner.
5. Humidification unit configuration, capacity and compatibility to space available.
6. Humidification equipment controllability and system control parameters.
7. Maintenance requirements and specifications.
8. Qualifications and capabilities of maintenance personnel to operate and service the equipment.
9. Humidification equipment noise characteristics and qualifications to meet system requirements or specifications.

10. Hazardous environment restrictions—pneumatic controls are mandatory if explosive environmental conditions can potentially develop.
11. Initial cost of humidification equipment and ensuing operating expense.

The foregoing factors should be evaluated and prioritized on a check list basis prior to the humidification equipment selection and finalization of system design.

HUMIDIFICATION SYSTEM LOCATION

The humidification unit must be located in the air handling system to provide maximum vaporization and optimum distribution. High temperature air and turbulence are desirable to achieve total vaporization. What is highly undesirable, however, is water formation in the air handling unit or ductwork caused by incomplete vaporization or condensation. Water formation on cold surfaces is detrimental in that it causes corrosion and erodes metal surfaces. Transient water droplets presents an even greater problem because it provides a favorable environment for bacteria and algae infiltration. Although algae formation is annoying, bacteria growth can potentially develop into a life threatening health hazard.

Based on the foregoing factors, it highly is recommended that humidifier emission units be located downstream from the fan and preheat/reheat coils. This presents the best possible psychrometric conditions and moisture absorption by the air stream. However, a note of caution is warranted: do not position the humidifier too close to the intakes of high efficiency filters. This positioning can result in the filters absorbing and retaining moisture. A minimum distance between filters and emission units should properly be approximately eight feet. Also considered good practice is to maintain at least an eight foot distance between the humidifier and a branch or split in the ductwork. This insures uniform moisture delivery throughout the system.

Commercial, Industrial Humidification

Despite general industry recommendations for humidifier location and application, the final authority in this matter must rest with the equipment manufacturer. All reputable manufacturers are sensitive to the proper application of their equipment. Therefore the authoritative formulation of specific instructions is given a high priority status. Before the instructions are finalized, input from all relevant branches of engineering and marketing are evaluated and integrated into the final, complete document.

If installers or application personnel have questions, doubts or if specialized problems in equipment adaptation to air distribution systems occur, always refer to the manufacturer's representative, application or design engineer. Never hesitate to pursue this avenue for authoritative information and assistance. Satisfactory results often depend on information only the manufacturer can provide.

On the job factors or circumstances may sometimes present conditions that require deviations from recommended and standard humidifier locations. A case in point is determining the most appropriate location for steam distribution manifolds. The preferred location is downstream from the fan and heat/reheat coils. If this is physically impractical, the manufacturer may suggest an upstream location as far from the run inlet as possible and at least three feet or more from a coil inlet and at least five feet from a temperature controller sensor.

There are two compelling reasons why a manufacturer's recommendations should be solicited and followed. First, recommendations by the manufacturer's representative are based on collective engineering expertise corroborated by both laboratory and field tests. Second, equipment installed in compliance with the manufacturer's recommendations and instructions fortifies and maintains the manufacturer's total performance and warranty responsibilities when the equipment is in operation.

HUMIDIFICATION SYSTEM CONTROL

The key humidity control is the humidistat. It senses humidity level generated in the air handler or the humidity conditions in occupancy zones and processing areas. Initially, humidity sensors were organic hygroscopic material such as human hair, membrane, animal horn, wood or biwood. Increase or decrease in humidity causes the hygroscopic material to expand (lengthen) and contract in a predictable manner. In the case of biwood (two layers of wood), the expansion and contraction characteristics differ for the two layers. This creates a warping action which actuates electrical contacts. These sensors were primarily used with on-off controllers, although human hair sensors have been utilized to actuate variable resistance or slide wire potentiometers. These applied as one leg of a wheatstone bridge circuit are capable of modulating or proportional control of humidity.

More recently, synthetic or nylon type sensors have been utilized. One type cellulose acetate butyrate (CAB) changes both physically and electrically according to humidity changes. Consequently, the electrical resistance change can provide a signal proportional to the humidity change for modulating performance. Also involving electrical resistance change are salt coated wire elements. Allied to the humidity sensor is the dew point sensor which uses optical and conductive means to sense moisture content and dew point formation. The electrical resistance change sensors are a significant break through for modulating circuitry as prior mechanical modulation humidistats required abnormally large devices. Large sensing elements were required to provide sufficient mechanical force to actuate slide wire potentiometers. Electrical resistance humidistats, on the other hand, eliminate mechanical error and size. In addition, the change in resistance signal provides an analog input to direct digital control systems for microcomputer control of hvac equipment or any building automation system. This is in keeping with the trend toward electronic and microelectronic circuitry.

Commercial, Industrial Humidification

Figure 5-9a illustrates the a basic humidity control for a heating and ventilating system. The space humidistat modulates a normally closed steam valve which supplies the humidifier located downstream from the fan. The solenoid air valve (SAV) in the diagram allows pneumatic control air to the steam valve as required by the humidistat. The SAV is wired in parallel with the fan. So, when the fan is inoperative, the SAV closes off pneumatic air pressure to the steam valve. The steam valve closes and

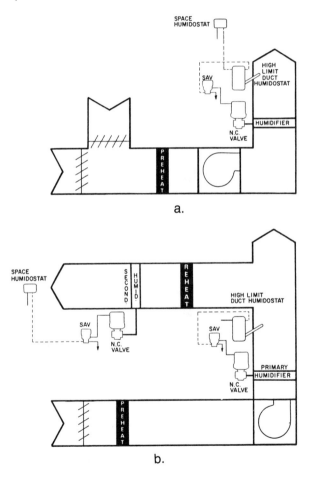

Figure 5-9. Basic humidity control for: a) a heating/ventilating system, b) large volume air conditioning systems — primary and secondary humidification.

prevents moisture emission without air flow. The high limit humidistat down-stream from the humidifier guards against excessive humidity which could create condensation problems.

Figure 5-9b illustrates a large volume air conditioning central fan system which utilizes 100% outdoor air. The primary humidifier can maintain space conditions up to 35% RH with 70°F dry bulb temperatures. The secondary humidifier, on the other hand, can boost humidity to a higher level for zones with greater requirements (e.g., 35% to 55% RH for computer rooms and hospital areas).

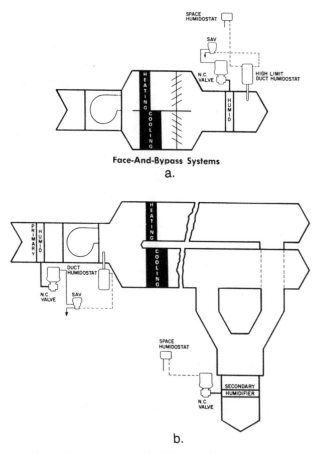

Figure 5-10. Humidifier application for: a) face-and bypass system, b) high velocity, double duct system.

Figure 5-10a is a humidifier application in a face and bypass central fan system. This system can provide heating, cooling or a combination of both to maintain a predetermined air temperature. The humidifier located down-stream from the damper section avails itself of the highest dry bulb air temperature. This maximizes moisture absorption in the air stream.

Figure 5-10b is an example of a high velocity, double duct central fan system. This system gives excellent total environmental control. Due to space constraints, the primary humidifier is usually located ahead of the inlet to the supply fan. The humidifier must maintain at least a five foot distance from the fan inlet. This primary humidifier supplies the basic requirements for the system. The duct humidistat senses humidity level down stream from the fan. Its set point must preclude condensation by accounting for the system dew point.

The secondary humidifier is installed in the discharge of a mixing box and ahead of the grille for each area or zone requiring a higher level of humidity. Although Figure 5-9 and -10 illustrate basic pneumatic controls, electro-mechanical controls are quite capable of providing the same functions.

In more sophisticated energy management and building automation systems, where the total internal environment is coordinated and controlled by a programmed microcomputer, humidification sensors provide input which the microcomputer analyzes. Once analyzed, the microcomputer initiates the logical output response. The microcomputer can operate as a "stand alone" unit in a selected area or zone and still receive central instructions or directions from a host computer. The host computer, for example, may initiate control functions to stop electrical loads. This load shedding could be for maximizing energy utilization or to prevent maximum electrical demand from being exceeded.

START-UP & TEST PROCEDURE

After a humidifier has been installed, the next step is the initial start-up and system check out. The manufacturer's start-up instructions, check list or both provide the recommended

Humidity Control

procedures and guide lines for this phase. Generally these involve:
1. Check of piping, water pressure and adequacy of drains.
2. Check of humidifier mounting—verify unit is secure and positioned according to manufacturer's specifications.
3. Check of electrical wiring—verify that wiring is in accordance with industry standards and in compliance with national electrical code and local codes.
4. Fan start-up (central fan systems).
5. Validation of control circuit—initiate a humidification cycle and check controls for proper response, sequencing, normal operation and limit functions.
6. Leak check air distribution system—also look for effective air delivery and condensation.
7. Make a capacity check—after a system normalizes, make instrumented readings of supply air and the humidified discharge air. Plot the readings on a psychrometric chart and analyze and document capacity and performance.

SERVICE & MAINTENANCE

The most common humidifier problems are water related. Even in geographic areas with minimal limestone content, water still has some mineral and chemical impurities. These impurities (contaminates) tend to accumulate and increase in concentration as vapor is generated. When these impurities solidify in a mass, they must be removed by periodic cleaning. In pan type humidifiers, timed automatic flush cycles can prolong the need for manual cleaning but will not totally eliminate it. Periodic inspection and cleaning should be incorporated in a documented maintenance procedure. A cursory inspection can be made after the equipment has been in operation for a week: this can the serve as a guide for future inspections.

With steam humidifiers, the steam boiler or generator must be treated with chemicals followed by mechanical cleaning or flushing. Generally this is the jurisdiction of the steam engineer. Nevertheless, maintenance personnel involved with humidifiers should make certain that boiler cleaning compounds have been thoroughly flushed out of the humidifier before start-up. Otherwise, objectionable odors can be introduced through the air distribution system.

Other problems related to fluid involve water feed or steam flow mechanisms. Impurities can affect these components by blocking or restricting fluid flow or preventing tight shut-off during an off-cycle or a limit control function.

Control problems can result from unauthorized tampering, misapplication or a malfunction. In perplexing cases, the problem can be quickly and systematically localized by circuit tracing and identification using a manufacturer's wiring diagram and a VOM (volt ohm meter).

Further, additional problems can involve condensation in the ductwork or condensation formation on windows or other cold surfaces. This means the cold surfaces are below the dew point of the humidified air. Psychrometric analysis and control readjustment is a solution for this problem.

A well-structured preventive maintenance program is the best insurance for a trouble free humidification and hvac system. This involves the development of a systematic check list and inspection procedure. All key hydronic, mechanical, and electrical components must be given periodic inspection and test, and the results accurately documented. With the advent of direct digital control systems and software, functions involving equipment application and maintenance, inspection, monitoring and even remedial operations can be accomplished from any remote location.

SUMMARY

Humidification is mandatory for many commercial and industrial storage, packaging, processing and production functions. It is downright essential for many institutional and human occupancy areas. Thus to fulfill its mission and justify its existence, humidification systems must incorporate established engineering procedures, psychrometric parameters and advanced technology in its design. When this is combined with systematic application, maintenance and service, the total environmental control loop is complete and functional.

SECTION 6

DEHUMIDIFICATION

An optimum humidity level is always desirable and, oftentimes, essential for indoor environments. In commercial/industrial processing, packaging and storage operations, humidity may have to conform to very close specifications. Humidity usually involves the addition of moisture—especially in the winter season. Nevertheless, moisture removal is also an important factor which requires well designed and properly selected equipment to meet load requirements. In the maintenance of optimum indoor environments, there is often a need for both humidification and dehumidification. This is especially true in the midwest area of the United States.

A high humidity level tends to interfere with the human body's ability to reject heat and regulate temperature by evaporation. Consequently, dehumidification becomes necessary. However, the process of dehumidifying an environment may create another problem. Removing moisture by lowering air temperature below the dew point temperature can lower the dry bulb temperature below a comfortable level. A reheat operation is required for this manner of moisture removal, and it may seem incongruous to summer air conditioning. Still, despite the additional equipment and operating cost, dehumidification is necessary in order to maintain design specifications.

PRODUCTION & PROCESSING

Success or failure in many production and processing operations depends on the maintenance of a precise degree of ambient air dryness. Specific examples include the production and processing of breakfast foods, soda crackers, instant coffee, powdery foods and potato chips to name a few. Each of these products, as well as many others, has a high affinity for moisture. If the surrounding air is too humid, the products absorb moisture and become soggy, rancid or degraded. The manufacturing of hard candy is another example where air moisture can create stickiness and handling problems. Drug production is another instance where humidity interferes with organic cultures and can deteriorate or even decompose tablets during the manufacturing process. In addition, air dryness is essential to quality control during the production of paper and many industrial chemicals.

To ensure proper functioning of production machinery operations, low dew point air is crucial. In other processes, condensation or frost formation can cause malfunctions or ruin product processing. Even in operations such as fluorescent tube manufacturing, dry air must be provided to shorten drying time.

Storage Requirements

In order to protect many products in storage from rust, corrosion, mildew, and damp rot, dry air is required. Hygroscopic material such as flour, sugar, synthetic materials or plastics are marketable only if moisture absorption is strictly controlled or eliminated. Generally, humidities less than 40% RH are recommended for storage operations.

Packaging Requirements

The control of dry air is essential for packaging hygroscopic products. If extraneous moisture is adsorbed and enclosed in packaging, even a small amount, the product can congeal and become rancid or stale. Moisture in air also affects the operation of packaging equipment or may interfere with the sealing process or adhesion of sealing material.

DEHUMIDIFICATION METHODS

Compression Method

The compression method entails mechanically raising the dew point and partial pressure of water vapor in the air's gas/vapor mixture then cooling and allowing it to re-expand to produce condensation and water rejection at higher air temperatures. This method is costly and requires: mechanical compression equipment, high horsepower operation and an after cooling step that utilizes some type of liquid coolant. For the most part, these requirements limit the use of this method to specialized low air volume applications.

Refrigeration Method

When air is cooled below its dew point, dehumidified air results. Consequently, by passing air over or through a refrigerated coil or surface, which is maintained below dew point, moisture condenses and can be rejected. The psychrometric chart in Figure 6-1 illustrates the process and determines the precise amount of moisture (pounds or grains) that can be extracted from a pound of air when cooled below its dew point.

Air at 85°DB, 50% RH passes through a cooling coil which reduces the air temperature to 55°DB and 100% RH. The initial dew point (DP) temperature was 64° and the final dew point temperature is 55°. Moisture had to condense, and the left hand humidity ratio scale determines the amount of condensation. Air at 85°DB and 50% RH has 0.0129 lbs/moisture or 90 gr/lb of dry air. Air at 55°DB, 100% RH has 0.009 lbs or 65 gr/lb of dry air. The difference (90 − 65) is equal to 25 grains.

If a factor of 10,000 cfm is specified for the calculation, the actual amount of moisture removal can be determined. The psychrometric chart value of 25 grains removed was for one pound of dry air. The psychrometric chart indicated the initial air conditions of 85°DB air at 50% RH also pinpointed the specific volume of 14 ft^3/lb of dry air. The specific density (lbs/

Humidity Control

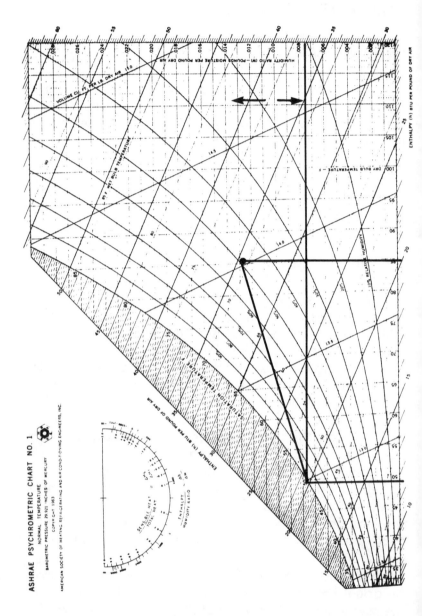

Figure 6-1. Psychrometric chart illustrating air at 85°DB, 50% RH cooled to 55°DB, 100% RH with a final dew point temperature of 55°.

Dehumidification

ft^3) is equivalent to the reciprocal of 14 or 0.07143. Multiplying 0.07143 by 10,000 ft^3 yields 714.3 lbs or the amount of air that passed through the cooling coil in one minute. The moisture removal per pound was 25 grains. Multiplied by 714.3 lbs reveals that a total of 17,857.5 grains of moisture was removed. Converting grains to pounds of moisture removed per minute involves dividing 17,857.5 by 7,000 which yields 2.55 lbs/minute or 153 lbs/hr.

The refrigeration capacity necessary to remove 153 lbs moisture/hr can also be computed by using the psychrometric chart. At the starting conditions of 85°DB and 50% RH, the wet bulb temperature is 70.8°WB; this indicates a h_1 total heat content per pound of dry air (h_1) of 34.6 Btu. At 55°DB, 100% RH, the wet bulb is also 55°. The total heat content at final condition (h_2) is equal to 23.3 Btu. Subsequently, the required heat removal is $h_1 - h_2$ or $34.6 - 23.3 = 11.3$ Btu.

By multiplying 11.3 Btu by the number of lbs air/minute and then by 60 reveals the Btu capacity/hr. Dividing that figure by 12,000 establishes the capacity in tons of refrigeration required to condense and remove 153 lbs moisture/hr. This calculation is as follows:

$$\frac{(h_1 - h_2) \text{lbs air/min.} \times 60}{12,000} = \text{tonage}$$

$$\frac{11.3 \times 714.3 \times 60}{12,000} = \frac{484,295.4}{12,000} = 40.36 \text{ ton of refrigeration}$$

If the previous example is a comfort cooling application, the air must be reheated from 55°F to a more acceptable dry bulb temperature. Plotting the new dry bulb temperature on the psychrometric chart results in a horizontal projection from the 55° line. The amount of reheat then is the difference in enthalpy between 55° and the new DB temperature multiplied by the lbs of air/hr.

The ability of cooling coils to dehumidify depends on its temperature below dew point and the coil surface area. In air conditioning, temperature below dew point is limited to about 40°F. Lower coil temperatures tend to create freezing problems. The coil surface area is the most essential variable when considering how to obtain maximum dehumidification. This variable is significantly effected by increasing the numbers of rows in a coil. Consequently, coil selection is an important capacity determinant and design consideration for direct expansion or chilled water dehumidification.

DESICCANT ADSORBENT DEHUMIDIFICATION

Dehumidifying by lowering air temperature below dew point is effective and practical for comfort cooling of occupied areas. A requirement for a low humidity condition coincides with the need for a lower dry bulb temperature. However, if dry air is the only requirement (such as in production and processing operations), then over-cooling the air may not be energy efficient.

One practical method of strictly producing dry air is the use of desiccant drying. A desiccant adsorbent is an inorganic material that has a high affinity for moisture. In adsorbing or rejecting relatively substantial amounts of moisture, the desiccant undergoes no physical or chemical change. Moreover, the chemical and physical stability exhibited by a desiccant allows it to be re-used or indefinitely re-cycled. A high quality household sponge, for example, has a high affinity for water and is capable of absorbing a large amount of water in its liquid state. Desiccants, on the other hand, adsorb water vapor from air. In addition, desiccants can remove liquid water from other liquids. A desiccant can remove water from liquid refrigerant in a refrigeration system for example. Some common desiccants include silica gel, activated alumina and molecular sieves.

Just as a household sponge is only useful if it retains its ability to absorb and reject liquids, an adsorbent desiccant only remains useful if it can adsorb and reject moisture in either liquid or vapor form. To rejuvenate a desiccant, it is subjected to heat in order to allow the desiccant to reject the moisture it contains.

Dehumidification

There are two types of desiccants—adsorbent and absorbant. The absorbent desiccant differs from the adsorbent type in that it changes either physically, chemically or both during a dehumidifying cycle. Absorbant desiccants may even undergo a phase change, such as from solid to liquid. Lithium chloride is a prime example of an absorbent desiccant. Selection of a desiccant for a given application depends on many factors. A list of desirable desiccant factors follows:
- high moisture retention capacity
- physical and chemical stability
- resistant to contamination and deterioration
- non-toxic, odorless, inflammable and non-corrosive
- capable of regeneration with practical and available methods and at readily attainable temperatures
- good vapor pressure characteristics—when regenerated, vapor pressure should be minimal
- commercially available at a reasonable cost

Most commercial desiccant type dehumidifiers use solid adsorbents. The adsorption process takes place on the surface of the adsorbent. The amount of moisture adsorbed is directly proportional to the available surface area of the adsorbent. The surface area is relatively large because adsorbents typically have a porous submicroscopic structure. In the adsorbent process, moisture molecules are pulled into the porous structure of a desiccant. This process is accelerated when a substantial vapor pressure difference exists between ambient air and the adsorbent. As vapor pressure equalizes, the adsorbent becomes saturated and attains equilibrium capacity. When this happens, the adsorbent must he rejuvenated. Factors which affect reaching equilibrium capacity include: air temperature, moisture content of the air and percentage of water adsorbed by the desiccant.

The adsorption process is reversible, so rejuvenation and re-use of adsorbents is possible. When vapor pressure of the adsorbed moisture becomes higher than the surrounding atmosphere, the desiccant rejects moisture. The moisture rejection is accomplished by heating the desiccant material. Temperatures in the range of 194° to 500°F (90° to 200°C) are required for the rejuvenation cycle. However, despite the rejuvenation, the desiccant still retains 5% to 6% of its moisture content. Any attempt to remove this residual moisture causes physical deterioration of the desiccant. If an adsorbent desiccant is vapor actuated and rejuvenated within its physical limitations, it can be cycled thousands of times before being expended.

Liquid Absorption

Liquid absorption equipment pumps a strong absorbent solution from a sump and sprays the solution over contactor coils. Air, which is to be dehumidified, is directed parallel to or in counter-flow through the coils. Either of these patterns creates intimate contact with the solution. The dehumidification rate then depends on temperature and vapor pressure difference between the absorbent and air. Dehumidification control is basically accomplished by controlling the temperature of the absorbent solution. A section of a dehumidifying chamber coolant coil facilitates this control. The coolant contained in the coil can be city water, chilled water or even a refrigerant.

Since an absorbent solution is continuously rejuvenated, a percentage of the absorbent flow from the contactor coil is diverted to a generator coil. The regenerator coil is frequently heated by steam. When exposed to a heated regenerator coil, moisture is driven from the absorbent to a scavenger air stream and rejected.

DEHUMIDIFICATION EQUIPMENT

Most solid adsorbent dehumidification equipment for commercial applications utilizes either silica gel, activated alumina or molecular sieves. A static dehumidification operation, which uses no forced air circulation, is also used in some cases. The air surrounding the desiccant in such a static operation is dried and dispersed by way of convection and vapor diffusion allowing higher humidity air to move forward for drying. A more effective method, however, is dynamic dehumidification. A dynamic system utilizes a forced air fan to move air through the desiccant bed and also through the regeneration chamber.

When air passes through a dehumidifier, the ratio of water adsorbed compared to the amount in the entering air is considered the **sorption efficiency**. When this ratio or efficiency begins to drop, due to desiccant saturation, the condition is referred to as the **breakpoint**. Consequently, the amount of water in the desiccant at this point is known as the **breakpoint capacity**. Breakpoint is the optimum time required to initiate the regeneration or rejuvenation cycle.

As heat is applied during the regeneration process, the dry bulb temperature of regenerated air rises rapidly to a certain point and then levels off as desorption of water takes place. When most of the water has been eliminated, a sharp dry bulb temperature increase occurs in the effluent or discharge air emanating from the desiccant. The period from initiation of regeneration to the pronounced increase in effluent dry bulb temperature is referred to as the **temperature-rise time**. Although regeneration may still continue, at a somewhat reduced rate, the end of the temperature-rise time is considered the ideal point at which to terminate regeneration and to initiate dehumidification.

As the desiccant starts adsorbing water from water vapor, latent heat is released while a change of state occurs. As pointed-out in Section 1, this requires a substantial amount of heat equivalent to approximately 1,000 Btu/lb. An additional quantity of heat is also released; this is known as the **heat of wetting** and results when a liquid and a solid surface come in direct contact with one another. The heat of wetting is substantial when adsorption starts, but it diminishes as the desiccant approaches saturation.

All the heat released in the adsorption process is called the **heat of sorption**. It increases the temperature of the air stream, the enclosure and the desiccant bed itself. If the desiccant bed temperature increases to 250°F (121°C), the elevated bed temperature can decrease the efficiency of the operation.

Other factors that affect desiccant bed efficiency include desiccant:
- type
- weight
- particle size
- density
- bed configuration
- bed depth and compactness
- pressure drop through the bed

Factors that affect the air to be dried include flow:
- rate
- temperature
- pressure
- moisture content
- time factor—duration of contact between the air and desiccant bed

The bed reactivation factors include:
- reactivation temperature
- amount and rate of heat added
- the air flow rate
- bed heat storage capacity
- insulation factor

Supplemental and additional factors that can affect and increase capacity are the use of pre-cooling and after-cooling coils. These can be direct expansion refrigerated coils or chilled water coils. These coils can provide initial moisture removal and temperature decrease of the outlet or discharge air.

ADSORPTION DEHUMIDIFICATION EQUIPMENT

Figure 6-2 is a schematic view of a rotary dehumidification unit. Rotary dehumidifiers have the highest capacity per unit of equipment volume. Its capacity and space efficiency account for its popularity.

A typical rotary unit has one or more desiccant beds which are subjected to a parallel air flow. This flow pattern allows one bed to dry air while the other bed is simultaneously being regenerated. In order to isolate the simultaneous drying

Dehumidification

and regeneration operations, the two air flows are separated by seals or baffles. The beds are physically rotated in periodic cycles which can range from ½ rotation/hour to as many as 6 rotations/hour. The dual bed unit provides continuous drying—one bed is always in the drying mode while the other is being rejuvenated. The single bed unit, on the other hand, utilizes intermittent drying since the single bed must be regenerated periodically.

In Figure 6-2 the wet air inlet consists of return or recirculated air which combines with a predetermined amount of fresh air. Most air handling and humidification systems require a source of outside or make-up air. The make-up air is used to meet ventilation code requirements as well as replace air for processes and exhaust operations. The recirculated, moist return air is routed towards the rotating desiccant bed while in the drying mode. Water vapor is then extracted from the mixed air stream, and the dry air that results is routed to

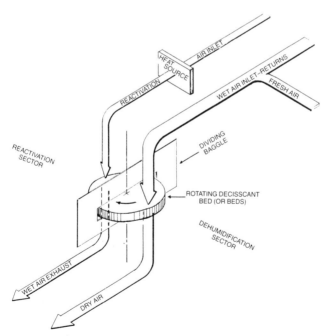

Figure 6-2. Typical rotary dehumidification unit fed by an outside air mix. (Used by permission—Bry-Air, Inc.)

Humidity Control

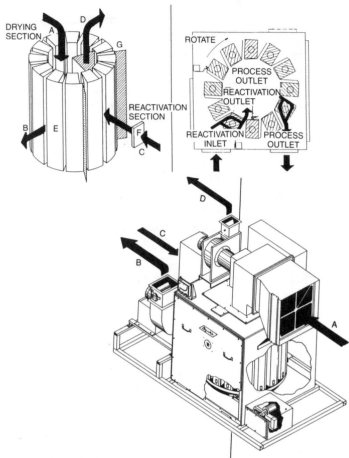

Figure 6-3. Rotary bed dehumidifier designed with vertical beds filled with desiccant. Air enters the unit through the process air inlet and moves through the desiccant beds in the chamber. The beds adsorb the water vapor, and the dehumidified air is delivered through the process outlet. The carousel of vertical beds rotates so one bed moves into the rejuvenation area where heated air enters and passes through the desiccant removing moisture which is discharged into the atmosphere. A seal between chambers prevents mixing of process air and the regeneration air streams. (Used by permission — Bry-Air, Inc.)

Dehumidification

the application area. A parallel air inlet at the heat source directs air to the regeneration desiccant bed where moisture is driven out and exhausted in the reactivation mode.

Figure 6-2 is an example of a recirculation system utilizing an outside air mix. Systems can be supplemented with an after-cooler down-stream from the drying chamber. The after-cooler involves a refrigerated or chilled water coil. This can be useful in removing sensible heat (derived from the change of state which converts latent heat in water vapor to sensible heat).

Refrigerated or a chilled water pre-cooler can also be used to refine the process. A pre-cooler may only use outside air which is very effective when a high outside air requirement exists. In other cases, a pre-cooler may use the total air mixture prior to it entering the desiccant drying chamber. Psychrometric analysis or the application and the capability of the dehumidifier are the most significant criteria to judge when selecting a system.

Although a rotary bed dehumidifier (Figure 6-3) is effective, other types of dehumidifiers which have stationary single- or dual-bed configurations can also be considered. The single-bed unit usually has a programmed timer and a valving arrangement which allows air through the desiccant bed for drying. When the programmed timer reverses the valving sequence, a rejuvenating heater is actuated and purged air is circulated through the bed. A dual-bed unit provides continuous drying. Its damper, fan and regeneration heater are all programmed to allow one bed to dehumidify while the other regenerates. Equipment for both absorbent desiccants (lithium chloride for example) and adsorbent types operate along the same lines.

DEHUMIDIFICATION LOAD CALCULATIONS AND SIZING

In dryer sizing and selection, moisture gains from all sources must be calculated and factored in to achieve an ac-

curate estimate. These load calculations must involve both internal and external load factors. Examples of internal loads include:
- occupancy load—latent heat moisture rejected by people
- product load
- permeation load
- infiltration load
- process load from products
- gas combustion vapor load
- exposed water surface load

Each of the loads listed must be handled by the dehumidifier. External loads, on the other hand, are exclusively either outside air (required to meet ventilation codes) or make-up air (for exhaust requirements) or both. As a rule, external loads can either be handled by pre-cooling coils or by increasing the dehumidifier capacity.

Load Calculation Example

This example involves a building where hard candy is produced. The physical facts concerning the building and its occupancy follow.
- area to be conditioned 60' × 42' × 16'
- outside design conditions 95° DB, 75° WB
- controlled space requirement 75°DB, 35% RH
- one 6' × 7' door, opens 6 times/hr
- occupants in work area = 10
- construction, 8" masonry
- make-up air 350 cfm (specified by owner)

Specific details of building construction and personnel activity are:
- door is adequately weather-stripped (standard construction)
- 10 occupants work continuously at moderate rate (each requires ventilating air)
- controlled space interior finished with two coats vapor barrier paint
- only the one opening to work space is provided
- vapor barrier exists under concrete floor

Dehumidification

Table 6-1. Latent heat dissipated by adult occupants. (Used by permission—Bry-Air, Inc.)

Dry Bulb Temperature	Occupants At Rest	Occupant Doing Light Physical Exertion *	Occupant Doing Heavy Physical Exertion **
60° F.	460 gr/hr	1300 gr/hr	1960 gr/hr
65° F.	530 gr/hr	1630 gr/hr	2400 gr/hr
70° F.	670 gr/hr	2060 gr/hr	2920 gr/hr
75° F.	900 gr/hr	2540 gr/hr	3450 gr/hr
80° F.	1180 gr/hr	3040 gr/hr	3950 gr/hr
85° F.	1525 gr/hr	3550 gr/hr	4450 gr/hr
90° F.	1870 gr/hr	4000 gr/hr	5000 gr/hr

*Examples—Waiters, dinner dancing, light factory assembly work.
**Examples—Factory machine operator, continuous dancing.

Occupancy or Population Load—As Table 6-1 indicates, the moisture rejection attributable to occupants depends upon the dry bulb temperature of the ambient air and the occupant's level of activity. In the example, 10 people are working in a 75°DB, 35% RH environment at a moderate rate. Their activity level is greater than light physical exertion but not as great as heavy physical exertion. Reading over from 75° in Table 6-1 suggests a mid-point interpolation between 2,540 and 3,450 gr/hr which is equivalent to 3,000 gr/hr. The population load for the example is:

10 people × 3,000 gph = 30,000 grains moisture/hr

Population load is assumed to be human population, but this can be an over-simplification in applications where animals and livestock are involved. Charted information on animals is practically non-existent, but if animal moisture rejection is significant, it is possible to make a calculation. This is done by weighing the drinking water of a representative animal over a given test period. The result is then presumed to be the moisture rejected during an estimated period of time.

Table 6-2. a) recommended design outside moisture level, b) F_1 factor for grain difference (Used by permission — Bry-Air, Inc.)

DESIGN		DESIGN	
Outside Wet Bulb	Specific Humidity	Outside Wet Bulb	Specific Humidity
°F	gr/lb	°F	gr/lb
81	150	75	128
80	145	74	123
79	140	73	120
78	137	72	115
77	135	71	110
76	132	70	107

a

Gr/lb Difference	F_1 Factor
35	1.1
40	1.11
50	1.35
60	1.58
70	1.82
80	2.05
90	2.29
100	2.52
110	2.76
120	2.99

b

Product Load — Facts given in the example summary do not indicate any significant moisture load. Still, there are many products introduced in a controlled area which may have substantial moisture content. ASHRAE and manufacturer's tables give the moisture content of many hygroscopic materials, and this can be used as a basis for product load determination. However, a better method is an on the job test. A pilot sample is placed in area which simulates actual field conditions; then the moisture loss of the sample is measured after a suitable time interval.

Dehumidification

Table 6-3. a) F_2 factor for space permeation, b) F_3 construction factors, and F_4 vapor barrier factors. (Used by permissionn — Bry-Air, Inc.)

a.

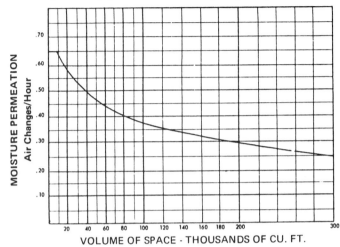

b.

F_3 FACTORS FOR CONSTRUCTION***	
Masonry or Frame construction	1.0
Sheet metal, steel welded	0.2
Module panel, caulked and sealed	0.5

F_4 FACTORS FOR VAPOR BARRIERS***	
Laminated mylar — metallic or polyeth. film	0.5
Two layers edge sealed moisture paper	0.67
Two coats vapor proof paint	0.75

***If the product of $F_3 \times F_4$ is less than 0.5, use 0.5. If the room is completely vapor proofed, with continuous vapor barrier under the floor (or of all-metal, welded material) the factor may be reduced to 0.2.

Permeation Load — Permeation is a load created by a low vapor pressure environment surrounded by an outside atmosphere of high vapor pressure. The resulting imbalance is so pervasive that it can easily travel through ordinary building construction. Even opposition from an air flow cannot prevent vapor pressure equalization. It can be reduced, however, by well selected and properly installed vapor barriers. Proper installation indicates that all laps and seams must be tightly closed and sealed. And precautions must be taken to prevent moisture from being trapped in wall and ceiling insulation. Despite preventive measures, permeation is still a significant factor to considered in any dehumidification load calculation.

Humidity Control

The formula for permeation load calculation follows:

$$\frac{V}{C} \times \Delta G \times F_1 \times F_2 \times F_3 \times F_4 = gr/hr$$

(Note, to convert to gr/min divide by 60)

Where:
   V = Volume of controlled space (ft^3)
   $C = 14$ = constant to convert ft^3 to lbs
      (used regardless of air density)
   ΔG = difference gr/lb outside air and
      gr/lb desired in controlled space
   F_1 = moisture difference factor (multiplier from Table 6-2b)
   F_2 = permeating factor (multiplier from Table 6-3a)
   F_3 = construction factor (Table 6-3b)
   F_4 = barrier factor (Table 6-3b)

The calculated permeation load for the example problem is:

$$\frac{40{,}320}{14} \times 82 \times 2.10 \times 0.49 \times 1.0 \times 0.75 = 182{,}250 \text{ gr/hr}$$

Where:
   $V = 60 \times 42 \times 16 = 40{,}320 \text{ ft}^3$
   $C = 14$ (specific volume dry air at 95°F)
   ΔG = outside design 75°WB = 128 gr/lb from Table 6-2a.
      Controlled space requirement of 75°DB, 35% RH yields
      46 gr/lb from psychrometric chart so 128 − 46 = 82 gr/lb
   $F_1 = 2.10$ moisture difference − 82 gr/lb
      (interpolated from Table 6-2b)
   $F_2 = 0.49$ (from Table 6-3a)
   $F_3 = 1.0$ factor for 8" masonry (from Table 6-3b)
   $F_4 = 0.75$ (from Table 6-3a) factor for 2 coats paint

Dehumidification

Infiltration Load—Infiltration of outside air through openings and cracks in building construction is a substantial load which can be calculated in several ways. A simplified, and perhaps over-simplified way is to presume how many times an air change occurs in a given period of time. This entails multiplying the building volume by the number of air charges or fraction thereof per hour. This volume in cubit feet can be converted to pounds by dividing by the specific volume (ft³/lb) of outside air at the selected design temperature. The solution is attained by multiplying pounds of air infiltration by the difference in moisture content of outside and inside air.

Openings can be intermittent such as personnel and service doors. Moisture that enters an environment when a door opens is a load that the drying equipment is required to handle. Consequently, it can be calculated as follows:

$$O_{hr} \times \frac{A}{C} \times \Delta G \times F_1 = gr$$

Where:
O_{hr} = number of times door opens/hr
 A = area of door opening (ft²)
 C = 7 constant (to convert ft³ to lbs)
ΔG = difference in specific humidity gr/lb between controlled space and adjacent space (refer to Table 6-2a for outside WB to determine adjacent space specific humidity)
F_1 = moisture difference (factor from Table 6-2b)

Fixed openings such as open windows and conveyors also need to be calculated.

$$\frac{A \times 300}{C \times D} \times \Delta G \times F_1 = gr/hr \text{ load through fixed opening}$$

Where:
 A = area of fixed opening in ft²
 300 = vapor velocity in ft/hr at 35 gr difference (experimental constant)
 C = 14 (constant to convert ft³ to lbs)
 D = depth of opening (expressed in ft)
ΔG = difference in gr/lb between wet space and drier space
F_1 = moisture difference (from Table 6-2b)

Humidity Control

For example: a conveyor opening of 2 ft², depth of opening 1.5 ft, moisture difference of 100 grains:

$$\frac{2 \times 300}{14 \times 1.5} \times 100 \times 2.52 = 7{,}210 \text{ gr/hr}$$

Infiltration through fixed openings can also be determined by measuring the lineal feet of crack. ASHRAE and manufacturers' tables provide the amount of leakage in ft³/hr for various types of windows and doors. When the amount of air in ft³ has been determined, the moisture load calculation is similar to the air change method.

$$O_{hr} \times \frac{A}{C} \times \Delta G \times F^1 = \text{gr/hr}$$

In the example problem, the infiltration load occurs because one 6'×7' door is opened an average of 6 times/hour. The actual calculation is:

$$6 \times \frac{42}{7} \times 82 \times 2.10 = 6{,}200 \text{gr/hr}$$

Where:
$O_{hr} = 6$
$A = 6 \times 7 = 42 \text{ ft}^2$
$C = 7$
$\Delta G = 82$
$F_1 = 2.10$

Product Process Loads — The example problem does not involve a process load. However, if steam or moisture is used in processing, it must be estimated and included in the load estimate.

Gas Combustion Vapor Load — The example problem does not include any gas-vapor load. However, moisture released in gas combustion must be included in load calculations as well. Some manufacturers provide tables to calculate moisture load produced in various gas combustion burners. If none are available, use the value of 650 grains of moisture/ft³ of natural gas/hr to formulate an estimate.

Dehumidification

Exposed Water Surface Load — Although the example problem did not involve any water surfaces, evaporation from water surfaces also adds to the moisture load. Where water surface evaporation does exist, the load calculation can be determined from tables provided by manufacturers. In the absence of these, assume that 20 grains of moisture/minute/ft² is heated by air movement of 100 ft/minute at a water temperature of 87°F.

Total Internal Load Example

The total internal moisture load for a building where hard candy is being produced consists of the following:
- permeation load   = 182,256 gr/hr
- door opening load =   6,200 gr/hr
- population load   =   30,000 gr/hr
- total internal load = 218,456 gr/hr

The resulting figure is represented in the schematic diagram illustrated in Figure 6-4.

The total internal load of 218,456 gr/hr must be delivered by the recirculated air to the dryer. The system fan must therefore have sufficient capacity to deliver the recirculated air in addition to any outside ventilation or make-up required by the system specifications.

The initial step is to calculate the fan capacity required for the recirculated air.

$$\frac{X}{C} \times (S - G) = \text{Total space moisture load (gr/min)}$$

Where:
X = delivery air rate from dryer to space (expressed as cfm)
$C = 14$ = specific volume constant (ft³ air at 95°f)
$S = 46$ = gr/lb moisture required for controlled space (without ventilation = inlet condition at dryer)
G = air leaving the dryer (expressed in gr/lb): Refer to Figure 6-5 — enter curve at 46 gr inlet moisture, intersect 75° inlet air temp curve at 8 gr/lb

Humidity Control

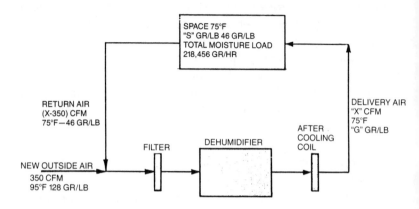

Figure 6-4. Schematic showing the dehumidification system considering the total moisture load of the building in the example. (Used by permission—Bry-Air, Inc.)

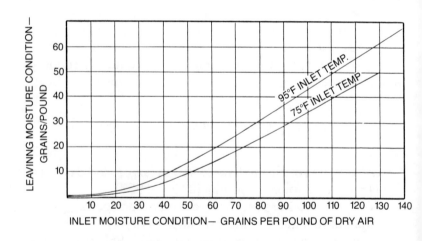

Figure 6-5. Typical performance chart. (Used by permission—Bry-Air, Inc.)

Dehumidification

Calculating:

$$\text{moisture load} = \frac{X}{14} \times (46-8) = \frac{219{,}000}{60} = 3{,}650 \text{ gr/min}$$

$$X = \frac{3{,}650 \times 14}{38} = 1{,}344 \text{ cfm}$$

The foregoing calculations established 1,344 cfm as the requirement for the internal moisture load. Add to this the 350 cfm required for make-up and ventilation air. This becomes approximately 1,700 cfm, and this can be the basis for an initial driver selection. A manufacturer's catalog might list a 2000 cfm dryer. The suitability of this dehumidifier can then be determined. With the 2,000 cfm capacity, 350 cfm is expended for the outside make-up air leaving 1,650 cfm for the recirculated air and the internal load. The air mixture can be tabulated on the basis of 350 cfm air containing 128 gr/lb and 1,650 cfm containing 46 gr/lb. The calculation follows:

$$\begin{array}{r} 350 \text{ cfm} \times 128 \text{ gr/lb} = 44{,}800 \\ 1{,}650 \text{ cfm} \times 46 \text{ gr/lb} = 75{,}900 \\ \hline 2{,}000 \text{ cfm} \hspace{2em} 120{,}700 \end{array}$$

Then,

$$\frac{120{,}700}{2{,}000} = 60.4 \text{ gr/lb}$$

Refer to Figure 6-5. This figure demonstrates that air entering the dryer at 60.4 gr/lb leaves the dryer at approximately 18 gr/lb. (Note: interpolate between 75° and 95° curves since the air is a mixture of 75° and 95° = 78.9°F).

$$\frac{X}{C} \times (S-G) = \text{total moisture pickup}$$

$$\frac{2{,}000}{14} \times (46-18) = 4{,}000 \text{ gr/min total removal capacity}$$

It appears, then, that the 2,000 cfm dehumidifier with 4,000 gr/min capacity is adequate enough to handle the 3,640 gr/min internal load. It should also have enough added capacity to handle the make-up air load. Nevertheless, some manufacturers add the outside ventilation or make-up air load to the internal load. This provides a more conservative sizing procedure which may result in selecting a larger unit as a safety factor. However, this can also result in an over-sized unit which can present control problems. Essentially, this becomes a judgement better left to the designer. The procedure recommended in this type of circumstance is a consultation with the manufacturer's application engineer.

START-UP, TESTING, SERVICE & MAINTENANCE

The procedure for dehumidifier start-up, test and maintenance is the same as that for humidifiers (as outlined in Section 5). Control operations of humidifiers and dehumidifiers are similar in that the most effective control is to simply turn them on or off as required. Dehumidifiers, however, can be an integral part of an automated process or building. The digital output and input of the dryer can be part of a microcomputer program. Although humidity sensors for humidification and dehumidification may be similar in many routine applications, there are specialized dehumidification processes which require highly sophisticated dew point sensors.

INDEX

absenteeism/humidity 97
absorption, liquid 136
adsorption
  dehumidification 138
air
  conditions 36
  flow 76
  movement 84
  noise 91
air change method 56
air mixture values 38
anti-syphoning device 76
area humidifier 116
ARI method 61
ARI standards 66-67
ASHRAE 12
  home 45
  method 59
atomizing humidifier 49, 111
auxiliary circuits 84
average house 57
back-syphoning 76
body heat 42
boiling point 28
Boyle's law 9
breakpoint 137
breakpoint capacity 137
British Thermal Unit (Btu) 1
Btuh 1
bypass humidifier 71
calorie 1
central fan calculation 103
centigrade scale 2-3
central system humidifier 66
change of state 4-7
Charles' law 9
charted load calculation 104
check-out procedure,
  humidifier 85
cleaning procedure,
  humidifier 86, 88-89

combustion chamber, furnace 75
commercial humidification 93
compression,
  dehumidification 131
condensation 23
condenser 65
conduction 3
control circuit, humidifier 84
control system, humidity 95
convection 3
corrosion 69
counterflow furnace 63
crack infiltration 53
crack method 53
cup humidifier 115
customer education 87
Dalton's law of partial
  pressures 10, 16
dehumidification 129
  equipment 136
  load calculation 142
density 27
desiccant bed 137-138
design factors 100
desorption 137
desiccant adsorbent
  dehumidification 134
desiccant reactivation 138
desiccants 135
dew point indicator 23
dew point temperature 22
downflow furnace 63
drain connection 76
dry bulb temperature 17
dusting, mineral 113
dusting, mineral 69
education 87
electric furnace 65
electrical connections 77-84
electrostatic charges 44
enclosed grid humidifier 115

Humidity Control

energy 1
enthalpy 27
enthalpy calculation 40
evaporation 42
evaporation rate 43
excess humidity 90-91
face and bypass system 124
Fahrenheit scale 2-3
fall-out, mineral 69
fly-wheel effect 4
furnace
  combustion chamber 75
  configurations 52
furnace,
  counterflow 63
  downflow 63
  horizontal 65
  upflow 63
furnishing protection 43
gas combustion
  vapor load 148
gas properties 9
health/humidity 43, 97
heat 1
  absorption 4
  flow 43
  intensity 2
  pumps 65
  rate 1
  rejector 65
  transfer 3
  transmission 3
  units 1
heat, body 42
heating system
  configurations 63-65
high velocity system 124
horizontal furnace 65
house classifications 57
humidification load 53, 62
humidification load
  calculation 99
humidifier
  check-out procedure 85
  cleaning procedure 86

control circuit 84
cycling 69
installation 73-85
limitations 69
location 120-121
mounting 71, 75
selection 66-73, 119
sizing 53
start-up procedure 85, 125
test procedure 125
wiring 78-84
humidifier,
  atomizing 49
  bypass 71
  central system 66
  fixed pad 70
  infrared 52, 71, 117-119
  pan type 47, 109
  rotating disc 51
  rotating drum 70
  steam 113-117
  wetted-element 50, 70
humidifiers 47, 109
humidifiers, atomizing 111
humidistat 79, 89
humidistat wiring 79-80
humidity 8, 25
humidity control 122-125
hydronic unit 65
hygroscopic material 44, 60, 97
immersion element 3
industrial humidification 93
infiltration rate 53
infrared
  humidifier 52, 71, 117-119
installation sites 68
installation, humidifier 73-85
internal moisture 60
jacketed dry-steam
  humidifier 115
latent heat 21, 42
latent heat of fusion 5
latent heat of vaporization 7
leveling, humidifier 75
lime scale 89

liquid absorption 136
lithium chloride 135
location, humidifier 120-121
loose house 57
low voltage source 78
maintenance 126
metabolism, body 42
mineral dusting 69, 113
moisture
  calculation 40
  removal 129
motor oiling 89
mounting
  location 71
  template 75
mounting, humidifier 75
muriatic acid 89
National electric code 77
no humidity check list 90
no humidity complaint 89, 94
nomograph load
  calculation 106-108
occupancy load 143
occupant comfort 42, 94
occupant heat dissipation 143
optimum humidity 129
overflow mechanism 75
packaging 130
pan type humidifier 47, 69, 109
percentage humidity 14
permeation load 145
population load 143
pound/grain conversion 14
product load 144
product/process load 148
production & processing
  dehumidification 129
productivity/humidity 97
psychrometers 19
  digital 20
  sling 20
psychrometric chart 13, 16, 62
  application 36
  construction 33
  values 17

quartz lamps, infrared 117
radiation 4
radiation, infrared 117
rate of evaporation 43
rational psychrometric formula 33
refrigeration 7
refrigeration,
  dehumidification 131
regulating devices 75
relative humidity 12, 21, 25
residential humidification 42
rotary dehumid. unit 138-140
rotating disc humidifier 51
rotating drum humidifier 70
rotating screen element 71
saddle valve 77
selection, humidifier 66-73, 119
self tapping valve 77
self-contained humidifier 116
sensible heat 5, 21
sensible heat ratio 29
service 126
  check list 92
  problems 87
  procedures 87
sizing dehumidification
  systems 141
sizing formula 57
solenoid valve 50
specific heat 2, 7
specific humidity 25
specific volume 27
square foot/cubic foot method 58
start-up procedure,
  humidifier 85, 125
static cling 98
static electricity 44, 99
steam humidifier 113-117
storage 130
system
  capacity 69
  control 122-125
test procedure 125
thermometer 1
thermostat setback 44

Humidity Control

tight house 57
total internal load 149
transformer 80
two speed blower operation 81-83
UL listing 77
Universal gas formula 9
upflow furnace 63
vapor
  content 21
  pressure 12-13, 28
vaporization 8

vinegar 88
water
  damage 75
  impurities 87-88
  minerals 50
water vapor migration 11
wet bulb
  depression 21
  temperature 19, 21
wetted-element humidifier 50, 70
wiring, humidifier 78

The solution to ice maker servicing is
CRYSTAL CLEAR
Understanding Commercial Ice Makers

$21.95

This comprehensive guide takes the mystery out of commercial ice makers.

Diversify your skills and expand your reputation as a service technician. This book gives you everything you need to become a successful ice maker repairman.

The roots of ice maker service, maintenance and problem-solving (as well as others) are presented in this guide to help you unlock your potential as an ice maker serviceman. You'll earn more money by exploring a whole new outlet for your skills and gain a more prestigious reputation as a knowledgeable and helpful service technician.

Many customers depend on their servicemen to educate and advise them on ice makers. Help your customers and yourself by learning the intricacies of ice makers.

Don't turn down ice machine repair work —
Order your copy of Understanding Commercial Ice Makers

SPECIALIZE IN CASCADE SERVICING

Expand Your Service Skills to Include the Lucrative Field of Cascade Service.

96 pages • $21.95

Cascade system servicing is not for beginners. However, if you have a solid working knowledge of refrigeration servicing, *Simplified Cascade System Servicing* shows you what you need to know to take the mystery out of servicing low temperature systems. It explains in detail the tools required and describes the service procedures used on typical cascade configurations. In addition, you'll learn to identify the various cascade configurations and how to service each one.

With this comprehensive guide to cascade servicing, you'll experience little difficulty in troubleshooting and servicing cascade system units of 25 cubic feet (1 horsepower) or less.

BNP Business News Publishing Company
P.O. Box 2600, Troy, MI 48007